Health and Safety Revision Guide

T0231075

This companion to the bestselling *Introduction to Health and Safety in Construction* is an essential revision aid for students preparing for their written assessments on the NEBOSH National Certificate in Construction Health and Safety.

Fully updated to the April 2015 specification, the revision guide provides complete coverage of the syllabus in bite-sized chunks, helping readers to learn and memorise the most important topics. Throughout the book, the guide links back to the *Introduction to Health and Safety in Construction* textbook, helping students to consolidate their learning.

▶ Small and portable, making it ideal for use anywhere: at home, in the classroom or on the move.
▶ Suggests useful tips on study and examination technique.
▶ Includes practice questions and answers based on NEBOSH exam questions.
▶ Everything you need for productive revision in one handy reference.

The *Health and Safety in Construction Revision Guide*, written by the renowned health and safety author and former NEBOSH Vice Chairman Ed Ferrett, will be an invaluable tool for students as they prepare for their NEBOSH exam and for their subsequent health and safety work.

Ed Ferrett was a member of the NEBOSH Board from 1995 until 2010 and Vice Chairman from 1999 to 2008. He has delivered many health and safety courses and has been a lecturer on NEBOSH courses for various course providers. He has been an External Examiner for an MSc course and BSc course in health and safety at two UK universities, a Reporting Inspector for Independent Further and Higher Education with the British Accreditation Council and a NEBOSH Ambassador. Ed is a Chartered Engineer and a Chartered Member of IOSH.

Health and Safety in Construction Revision Guide

for the NEBOSH National Certificate in Construction Health and Safety

Second edition

Ed Ferrett

Routledge
Taylor & Francis Group

LONDON AND NEW YORK

First published 2012 by Routledge

This edition published 2016
by Routledge
2 Park Square, Milton Park, Abingdon, Oxon OX14 4RN

and by Routledge
711 Third Avenue, New York, NY 10017

Routledge is an imprint of the Taylor & Francis Group, an informa business

British Library Cataloguing-in-Publication Data
A catalogue record for this book is available from the British Library

Library of Congress Cataloging-in-Publication Data
A catalogue record has been requested for this book.

ISBN: 978-1-138-91679-1 (pbk)
ISBN: 978-1-315-68938-8 (ebk)

Typeset in Univers LT by
Servis Filmsetting Ltd, Stockport, Cheshire

Contents

Contents

Preface

Welcome to the latest *Health and Safety in Construction Revision Guide*, for the NEBOSH National Certificate in Construction Health and Safety. The guide has been designed to be used together with the NEBOSH National General Certificate syllabus and the textbook *Introduction to Health and Safety in Construction* by Hughes and Ferrett. The guide gives only a basic summary of the NEBOSH National Certificate in Construction Health and Safety course, and a full explanation of all the topics is given in the textbook.

The Revision Guide has the following features:

▶ It follows the latest NEBOSH National Certificate in Construction Health and Safety syllabus;
▶ Revision notes for each of the elements of the two units – NGC1 (Management of health and safety) and NCC1 (Managing and controlling hazards in construction activities);
▶ A summary of the learning outcomes and key points is given for each element;
▶ Important diagrams are included to help revision;
▶ There is a final section to advise on the preparation for the examinations and some specimen answers to both long and short answer questions based on recent NEBOSH examination papers.

The Revision Guide will also be useful to those who have specific health and safety responsibilities in their jobs, such as Construction Site Managers and those who are studying on other courses related to building and construction that include important health and safety elements, for example HNC, HND and degree courses in Building, Architecture, Construction, Building Services and Energy.

The compact size of the Revision Guide ensures that it can be easily carried and used for revision at any time or place. It should be used throughout the NEBOSH course alongside the textbook and course handouts.

Good luck with your studies.

List of principal abbreviations

ACM	Asbestos containing material
ACoP	Approved Code of Practice
AFFF	Aqueous film forming foam
CDM	Construction (Design and Management) Regulations (UK)
CHIP	Chemicals (Hazardous Information and Packing for Supply) Regulations
CITB	Construction Industry Training Board
CLP	Classification Packaging and Labelling Regulation
COSHH	Control of Substances Hazardous to Health Regulations (UK)
dB	Decibels
dB(A)	Decibel (A-weighted)
dB(C)	Decibel (C-weighted)
DSE	Display screen equipment
EPA	Environmental Protection Act 1990
EU	European Union
FOPS	Falling-object protective structure(s)
HAV	Hand–arm vibration
HSE	Health and Safety Executive
HSG	Health and Safety Guidance
HSW Act	Health and Safety at Work etc. Act 1974 (UK)
ILO	International Labour Organization
INDG	Industry Guidance
IOSH	Institution of Occupational Safety and Health
ISO	International Organization for Standardization
Leq	Equivalent continuous sound level
LEV	Local exhaust ventilation

LOLER	Lifting Operations and Lifting Equipment Regulations (UK)
LTEL	Long-term exposure limit
MEL	Maximum exposure limit
MHSWR	Management of Health and Safety at Work Regulations (UK)
MIRA	Motor Industry Research Association
MSD	Musculoskeletal disorder
NEBOSH	National Examination Board in Occupational Safety and Health
OEL	Occupational exposure limit
OES	Occupational exposure standard
OSH	Occupational Safety and Health
PPE	Personal protective equipment
ppm	Parts per million
PUWER	Provision and Use of Work Equipment Regulations (UK)
RCD	Residual current device
RIDDOR	Reporting of Injuries, Diseases and Dangerous Occurrences Regulations (UK)
ROES	Representative(s) of Employee Safety
ROPS	Roll-over Protective Structure(s)
RPE	Respiratory protective equipment
RRFSO	Regulatory Reform Fire Safety Order (UK)
SPL	Sound pressure level
STEL	Short-term exposure limit
TWA	Time-weighted average
WAHR	Work at Height Regulations (UK)
WBV	Whole-body vibration
WEL	Workplace exposure limit
WRULD	Work-related upper limb disorder

Management of Health and Safety

1.1

Foundations in health and safety

Learning outcomes	
Outline the scope and nature of occupational health and safety	
Explain the moral, legal and financial reasons for promoting good standards of health and safety in the workplace	
Explain the legal framework for the regulation of health and safety, including sources and types of law	
Explain the scope, duties and offences of employers, managers, employees and others under the Health and Safety at Work etc. Act 1974 (HSW Act)	
Explain the scope, duties and offences of employers, managers, employees and others under the Management of Health and Safety at Work Regulations	
Outline the legal and organisational health and safety roles and responsibilities of clients and their contractors	
Outline the principles of assessing and managing contractors	

🗝 Key revision points

The definitions of hazard, risk, civil law, criminal law, common law and statute law ☐

The business case for health and safety (direct, indirect, insured and uninsured costs) ☐

The employer's duty of care and other common law and statutory duties ☐

Criminal offences and defences ☐

Civil liabilities and defences (particularly negligence) ☐

The legal framework – the Health and Safety at Work Act, the Management of Health and Safety at Work Regulations, absolute and qualified duties ☐

The powers of the enforcement officer, health and safety offences, including corporate manslaughter and corporate homicide, and penalties ☐

The role and functions of other external agencies ☐

The health and safety responsibilities and duties of employers to their employees and others affected by their undertaking, such as contractors, general public, visitors, patients and students ☐

The health and safety responsibilities of directors, managers, supervisors, employees and the self-employed ☐

Duties and responsibilities of manufacturers and others in the supply chain ☐

The duties and responsibilities between client and contractor to ensure a high standard of health and safety during the contract ☐

| The legal duties under the Construction (Design and Management) Regulations | ☐ |
| The principles of selecting and managing contractors | ☐ |

Definitions

▶ Accident – an unplanned event that results in damage, loss or harm
▶ Hazard – the potential of something to cause harm
▶ Risk – the likelihood of something to cause harm
▶ Civil law – duties of individuals to each other
▶ Criminal law – duties of individuals to the State
▶ Welfare – provision of facilities to maintain health and well-being of people in the workplace (e.g. washing, sanitary and first-aid)
▶ Residual risk – remaining risks after controls applied
▶ Near miss – any incident that could have resulted in an accident
▶ Dangerous occurrence – a near miss that could have led to serious injury or loss of life
▶ Common law – law based on court judgements
▶ Statute law – law based on Acts of Parliament.

Reasons for good health and safety management

Moral reasons

Need to provide a reasonable standard of care and ethical reasons to reduce:

▶ accident rates; and
▶ industrial disease and ill-health rates.

Legal reasons

Employers have a duty to take reasonable care of workers. Poor management can lead to:

- ▶ prosecutions; and
- ▶ civil actions – compensation claims.

Financial reasons

Poor health and safety management can lead to:

- ▶ direct; and
- ▶ indirect costs.

Good health and safety management can lead to:

- ▶ a more highly motivated workforce resulting in an improvement in the rate of production and product quality; and
- ▶ an improved image and reputation of the organisation with its various stakeholders.

Social reasons

These include:

- ▶ societal expectation of good standards of health and safety; and
- ▶ duty of care (details under 'Aspects of civil law' below).

Costs of accidents and ill-health

Direct costs

Directly related to the accident and may be insured or uninsured.

Insured direct costs normally include:

- ▶ claims on employer's and public liability insurance;
- ▶ damage to buildings, equipment or vehicles; and
- ▶ any attributable production and/or general business loss.

Uninsured direct costs include:

- ▶ fines resulting from prosecution by the enforcement authority;
- ▶ sick pay;

6

▶ some damage to product, equipment, vehicles or process not directly attributable to the accident (e.g. caused by replacement staff);

▶ increases in insurance premiums resulting from the accident;

▶ any compensation not covered by the insurance policy due to an excess agreed between the employer and the insurance company; and

▶ legal representation following any compensation claim.

Indirect costs

Costs which may not be directly attributable to the accident but may result from a series of accidents.

Insured indirect costs include:

▶ a cumulative business loss;

▶ product or process liability claims; and

▶ recruitment of replacement staff.

Uninsured indirect costs include:

▶ loss of goodwill and a poor corporate image;

▶ accident investigation time and any subsequent remedial action required;

▶ production delays;

▶ extra overtime payments;

▶ lost time for other employees, such as a first-aider, who attend to the needs of the injured person;

▶ the recruitment and training of replacement staff;

▶ additional administration time incurred;

▶ first-aid provision and training; and

▶ lower employee morale, possibly leading to reduced productivity.

Some of these items, such as business loss, may be uninsurable or too prohibitively expensive to insure. Therefore, insurance policies can never cover all of the costs of an accident or disease, because either some items are not covered by the policy or the insurance excess is greater than the particular item cost.

Employers' liability compulsory insurance

▶ legal requirement for all employers;
▶ covers the employer's liability in the event of accidents and work-related ill-health to employees and others who may be affected by their operations;
▶ ensures that any employee who successfully sues his/her employer following an accident is assured of receiving compensation irrespective of the financial position of the employer; and
▶ made available either by display or electronically at each place of business.

Legal framework for health and safety

Sub-divisions of law	
Criminal law	**Civil law**
▶ enforced by the State to punish individuals (and/or organisations)	▶ disputes between individuals (and/or organisations) to address a civil wrong (tort)
▶ individual is prosecuted by an Agency of the State (e.g. Police, HSE, Local Authorities, Fire Authority)	▶ individual(s) and/or organisations are sued
▶ individual(s) guilty or not guilty	▶ individual(s) are liable or not liable
▶ courts can impose fine or imprisonment	▶ courts can award compensation and costs
▶ proof 'beyond reasonable doubt'	▶ proof based on 'balance of probabilities'
▶ cannot insure against punishment	▶ employers must insure against civil actions (**Employers' Liability Insurance**)

Sources of Law	
Common Law	**Statute Law**
▶ based on judgements made by judges in courts	▶ law laid down by Acts of Parliament
▶ generally courts bound by earlier judgements (precedents)	▶ Health and Safety at Work Act 1974
▶ lower courts follow judgements of higher courts	▶ specific duties mainly in Regulations or Statutory Instruments
▶ in health and safety definitions of negligence, duties of care and terms such as 'practicable' and 'reasonably practicable' are based on common law judgements	▶ takes precedence over common law

Sub-divisions and sources of law

Figure 1 shows the relationship between the sub-divisions and sources of law.

Sources of Law:	Sub-divisions of Law:
Common Law	Criminal Law
Statute Law	Civil Law

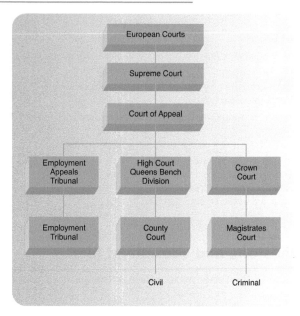

Figure 2 shows the court system for health and safety in England and Wales (NB European courts advise the Supreme Court).

Aspects of criminal law

The main prosecuting authorities in the UK are:

▶ the Crown Prosecution Service (CPS) in England and Wales;
▶ the Procurator Fiscal in Scotland; and
▶ the Public Prosecution Service for Northern Ireland (PPSNI) in Northern Ireland.

There are two types of criminal offence:

▶ Summary – minor offences, tried without a jury only in a Magistrates Court (or Sheriff Court in Scotland).
▶ Indictable – most serious offences are called indictable offences and tried only by the Crown Court (or the High Court of Justiciary in Scotland).

Comparison between English and Scottish courts

	Lower court	Higher court*
England and Wales	Magistrates	Crown
Scotland	Sheriff	High Court of Justiciary
Cases	Summary	Indictable

Note

* Higher court deals with some summary cases referred to it by the lower court.

Aspects of civil law

Employer's common law duty of care

The employer has a duty of care to each of his/her employees. This duty cannot be assigned to another person. The duty of care falls into five categories and the employer must provide:

▶ a safe place of work, including access and egress;
▶ safe plant and equipment;
▶ a safe system of work;
▶ safe and competent fellow employees; and
▶ adequate levels of supervision, information, instruction and training.

Employees have a right not to be harmed in any way by their work and they are expected to take reasonable care in their workplace.

Negligence

▶ lack of reasonable care or conduct resulting in injury, damage or loss;
▶ it must be reasonably foreseeable that the acts or omissions could result in injury.

Negligence claims must be made within a set time.

Defences against negligence claims	Partial defences against negligence claims
▶ A duty of care was not owed;	▶ Contributory negligence – employee contributed to the negligent act;
▶ There was no breach of the duty of care; and	▶ Volenti non fit injuria – the risk was willingly accepted by the employee;
▶ Any breach of a duty of care did not result in the specific injury, disease, damage and/or loss suffered.	▶ Acts of God, riot, terrorist event, etc.; ▶ a disagreement on the facts of the case.

Vicarious liability

When the defendant is acting in the normal course of his/her employment during the alleged incident, the defence of the action is transferred to his/her employer.

Tort of breach of statutory duty

This tort allows a person to seek compensation using a civil action following an accident or illness that resulted from a breach of statutory duty.

However, breaches of statutory duty, as described by regulations made under the HSW Act, only give rise to civil liability in cases concerning pregnant, or new or expectant mothers.

These employees may sue their employer both for negligence and the breach of a statutory duty. Such an action is termed a **double-barrelled action**.

Levels of statutory duty

Absolute	Practicable	Reasonably practicable
The employer MUST comply with the law. Regulations use the verbs 'must' or 'shall'	The employer must comply if it is technically possible. Difficulty, inconvenience or cost cannot be taken into account	If the risk is small compared with the cost, time and effort required to further reduce the risk, then no action is required

The legal framework for the regulation of health and safety

Influence of the European Union (EU)

- ▶ Role is to harmonise health and safety standards across member states
- ▶ Powers of EU in health and safety mainly derived from the Treaty of Rome (1957) and the Single European Act (1986)
- ▶ Article 95A (was 100A) – health and safety standards for plant and equipment
- ▶ Article 138A (was 118A) – minimum health and safety standards in employment
- ▶ European Directives set out the specific minimum aims of a given area of health and safety
- ▶ European Directives must be incorporated into the national law of all member states
- ▶ First introduction into UK law was in 1992 with the 'European Six Pack'
- ▶ European Court of Justice rules on interpretation of EU law
- ▶ European Court of Human Rights (covers a wider area than just the EU) interprets the European Human Rights Convention.

Health and Safety at Work Act 1974

Background topics

The Health and Safety at Work Act (HSW Act) was introduced in 1974 and is a criminal law based on the recommendations of the Robens Report.

Main recommendations of the Robens Report

1. A single Act to cover all workers containing general duties
2. The Act should cover everyone affected by the employer's undertaking
3. Emphasis on management of health and safety, including training and supervision

4 Encouragement of employee involvement in accident prevention
5 Enforcement should be targeted at 'self-regulation' rather than prosecution.

The HSW Act is an **Enabling Act**, allowing regulations to be produced by the HSE on behalf of the Secretary of State without further Acts of Parliament being required.

The HSW Act contains mainly general duties, with specific duties defined in Regulations.

Role and activities of the HSE

▶ advises on the development of regulations;
▶ enforces health and safety regulations;
▶ provides information to organisations (ACoPs, guidance notes, leaflets and other publications, accident and ill-health statistics);
▶ launches national health and safety campaigns on specific topics;
▶ undertakes accident and other investigations;
▶ offers advice to employers and others on statutory duties;
▶ takes enforcement action;
▶ instigates criminal proceedings and publicises organisations that receive enforcement notices.

Regulations

▶ state the law
▶ often implement EU Directives
▶ breaches are criminal offences possibly leading to enforcement action
▶ describe the minimum health and safety standards that need to be achieved
▶ usually apply across all organisations but sometimes apply to specific industries.

Approved Codes of Practice (ACoPs)

▶ supplementary practical interpretation of regulations that gives more detail on the regulatory requirements

▶ special legal status – quasi-legal because it may be possible to comply with regulations by some other more effective practice

▶ ACoPs are legally binding if the particular regulations indicate that they are or if they are quoted in an Enforcement Notice.

Guidance

▶ two forms of guidance – legal and best practice, both available as HSE publications

▶ not legally binding

▶ gives more informative and practical advice than ACoPs

▶ also often available as British Standards and as industrial or trade guidance.

The Health and Safety at Work Act

Section 2 Duties of employers to employees

General duty – ensure, as far as is reasonably practicable, the health, safety and welfare of all employees.

Specific duties:

▶ safe plant and systems of work;

▶ safe use, handling, transport and storage of substances and articles;

▶ provision of information, instruction, training and supervision;

▶ safe place of work, access and egress;

▶ safe working environment with adequate welfare facilities;

▶ a written safety policy together with organisational and other arrangements (if five or more employees); and

▶ consultation with safety representatives and formation of safety committees where there are recognised trade unions.

Section 3 Duties of employers to others affected by their undertaking

▶ 'others' could include contractors, general public, visitors, patients, students.

Section 4 Duties of landlords or owners

▶ safe access and egress for those using the premises; and
▶ buildings and any equipment supplied with them are safe.

Section 6 Duties of suppliers

Suppliers (including designers) of articles and substances for use at work to ensure, as far as is reasonably practicable, that

▶ articles are designed and constructed to be safe and without risk to health at all times when they are being set, cleaned, used and maintained;
▶ substances are similarly safe and without risk to health when being used, handled, stored or transported;
▶ arrange, where necessary, for suitable testing and examination; and
▶ supply suitable safety information and any revisions to customers.

Section 7 Duties of employees

▶ take care for the health and safety of themselves and others who may be affected by their acts or omissions; and
▶ cooperate with their employer, as far as is necessary, to ensure compliance with any statutory health and safety duty.

Section 8

No person to misuse anything provided for health, safety or welfare purposes.

Section 9

Employees cannot be charged for health and safety requirements.

Enforcement of the Act

Fee for Intervention (FFI)

The Fee for Intervention (FFI) scheme places a duty on HSE to recover its costs for carrying out its regulatory functions from those found to be in **material breach** of health and safety law.

Written notification by an inspector of a material breach includes the following information:

▶ the law that the inspector's opinion relates to;
▶ the reasons for their opinion; and
▶ notification that a fee is payable to HSE.

Powers of inspectors

▶ Enter premises at any reasonable time, accompanied by a police officer, if necessary;
▶ examine, investigate and require the premises to be left undisturbed;
▶ access to all records and other relevant documents;
▶ take samples, photographs and, if necessary, dismantle and remove equipment or substances;
▶ seize, destroy or render harmless any substance or article;
▶ take statements;
▶ issue improvement and prohibition notices and, possibly, a formal caution; and
▶ initiate prosecutions.

Enforcement notices

Improvement notice

Issued for a specific breach of the law.

Appeal within 21 days to an Employment Tribunal – notice suspended until appeal is heard or withdrawn.

17

Prohibition notice

Issued to halt an activity that could lead to serious injury.

Appeal within 21 days to an Employment Tribunal – notice is not suspended.

Deferred prohibition notice – stops the work activity within a specified time limit.

Penalties	
Summary offences	**Indictable offences**
▶ unlimited fines for all health and safety offences	▶ unlimited fines for all health and safety offences
▶ up to 12 months' imprisonment for health and safety offences*	▶ up to 2 years' imprisonment for all health and safety offences
▶ up to 5 years' disqualification for convicted directors	▶ up to 15 years' disqualification for convicted directors

Note

* The sentencing option of 12 months applies in Scotland but only applies in England and Wales when Section 154(1) of the Criminal Justice Act 2003 is enacted.

Work-related deaths

▶ Investigated by police initially to ascertain whether a charge of manslaughter (culpable homicide in Scotland) or corporate manslaughter is appropriate.

▶ If there are no such charges, the HSE or Local Authority continues the investigation.

The Corporate Manslaughter and Corporate Homicide Act

An organisation will have committed the new offence if:

▶ it owes a duty of care to another person in defined circumstances;

▶ there is a management failure by its senior managers; and

▶ it is judged that its actions or inaction amount to a gross breach of that duty resulting in a person's death.

The health and safety duties relevant to the Act are:

▶ employer and occupier duties including the provision of safe systems of work and training on any equipment used; and
▶ duties connected with:
 ▷ the supply of goods and services to customers;
 ▷ the operation of any activity on a commercial basis;
 ▷ any construction and maintenance work; and
 ▷ the use or storage of plant, vehicles or any other item.

A breach of duty is gross if the organisation's management of health and safety falls far below that which would reasonably be expected.

On conviction, the offence is punishable by an unlimited fine and the courts will be able to make:

▶ a remedial order requiring the organisation to take steps to remedy the management failure concerned;
▶ a publicity order requiring the organisation to publicise details of its conviction and fine; and
▶ an order to pay the costs of the prosecution and possibly compensation for bereavement and funeral expenses.

The publicity order requires an organisation convicted of corporate manslaughter to publicise:

▶ the fact that it has been convicted;
▶ the particulars of the offence;
▶ the amount of any fine;
▶ the terms of any remedial order.

Employer duties and responsibilities

Under the HSW Act, the employer has a duty to safeguard the health, safety and welfare at work of:

▶ employees;
▶ other workers – agency, temporary or casual;
▶ trainees;
▶ contractors;
▶ visitors; and
▶ neighbours and the general public.

19

Key actions required of the employer

▶ ensure the availability of competent advice on health and safety matters;

▶ obtain current Employers' Liability insurance and display the certificate;

▶ compile a health and safety policy and ensure that an adequate health and safety management system is in place;

▶ ensure that risk assessments of all work activities are undertaken and any required controls are put in place;

▶ provide the workforce with health and safety information and training;

▶ provide adequate welfare facilities;

▶ consult the workforce on health and safety issues;

▶ report and investigate some accidents, diseases and dangerous occurrences; and

▶ display prominently the health and safety law poster (or supply workers with the appropriate leaflet).

The Working Time Regulations set a working limit averaging 48 hours a week (over a 17-week rolling reference period). Young workers must not work more than 8 hours per day and 40 hours per week. Night workers are limited to an average of 8 hours' work in a 24-hour period.

Other employer responsibilities

Visitors and the general public

Possible hazards	Possible controls
▶ unfamiliarity with the workplace processes	▶ visitor identification (use of badges)
▶ lack of knowledge of site layout	▶ routine signing in and out
▶ unfamiliarity with emergency procedures	▶ escorted by a member of staff
▶ inappropriate personal protective equipment	▶ provision of information on hazards and emergency procedures
▶ inadequate or unsigned walkways	▶ given explicit site rules (wearing of PPE)
▶ added vulnerability if young or disabled visitors	▶ clear marking of walkways

For **night workers**, employers should:

▶ determine the normal working time for night workers;

▶ if the working time is more than 8 hours per day on average, determine whether the amount of hours can be reduced;

▶ develop an appropriate health assessment and method of making health checks;

▶ ensure that proper records of night workers are maintained, including details of health assessments; and

▶ ensure that night workers are not involved in work which is particularly hazardous.

Duties of managers and supervisors

Managers should:

▶ familiarise themselves with the health and safety management system of the organisation;

▶ ensure that there is an adequate and appropriate level of supervision for all workers;

▶ ensure that supervisors are aware of:

 ▷ the health and safety standards of the organisation;

 ▷ the specific hazards within their area of supervision;

 ▷ the need to set a good example on health and safety issues;

 ▷ the need to monitor the health and safety performance of their workforce; and

 ▷ the training needs, including induction training, of their workforce;

▶ ensure that sufficient resources are available to allow tasks to be completed safely and without risks to health; and

▶ communicate to the Chief Executive and the Board the adherence or otherwise of the health and safety standards agreed by the Board.

Supervisors should lead by example in health and safety matters (such as the wearing of PPE) and ensure that:

▶ workers in their department understand the risks associated with their workplace and the measures available to control them;

▶ the risk control measures are up to date and are being properly used, maintained and monitored;

▶ particular attention is paid to new, inexperienced or young people and those whose first language is not English;

▶ workers are encouraged to raise concerns over any shortcomings in health and safety provision; and

▶ arrangements are in place to supervise the work of contractors.

Duties and responsibilities of employees and others

Employees and agency workers

Under the HSW Act, employees and agency workers have specific duties:

▶ take care for the health and safety of themselves and others who may be affected by their acts or omissions;

▶ cooperate with their employer, as far as is necessary, to ensure compliance with any statutory health and safety duty; and

▶ not to interfere with or deliberately misuse anything provided, in accordance with health and safety legislation, to further health and safety at work.

Self-employed

Under the HSW Act the self-employed are:

▶ responsible for their own health and safety; and

▶ responsible for ensuring that others who may be affected by their undertaking are not exposed to risks to their health and safety.

In 2015, self-employed persons were exempted from the general duty in the HSW Act in respect of themselves and other persons (not being their employees), except those undertaking activities on a prescribed list defined by one of the following four criteria:

▶ there are high numbers of self-employed in a particular industry and high rates of injuries and/or fatalities (e.g. agriculture);

- there is a significant risk to members of the public (e.g. fairgrounds);
- there is the potential for mass fatalities (e.g. explosives); or
- there is a European obligation to retain the general duty on self-employed persons (e.g. construction – Council Directive 92/57/EEC imposes duties on the self-employed for safety and health requirements at temporary or mobile construction sites).

The occupational sectors for which no exemption was proposed included: agricultural activities; construction; quarries; mining; diving; sites that are subject to the Control of Major Accident Hazards (COMAH) Regulations; offshore activities; nuclear installations; explosives; and gas-fitting and installation.

Persons in control of premises

The duty of 'Persons in control of **non-domestic** premises' under Section 4 of the HSW extends to:

- people entering the premises to work
- people entering the premises to use machinery or equipment
- access to and exit from the premises
- corridors, stairs, lifts and storage areas.

Duties of manufacturers and others in the supply chain

Section 6 of the HSW Act (covered earlier) requires persons who design, manufacture, import or supply any article or substance for use at work to ensure, so far as is reasonably practicable, that it is safe and without risk to health, and employers should be provided with information on the safe use, dismantling and disposal of the articles and substances and given revised information should a subsequent serious risk become known.

Importers have a duty to ensure articles or substances comply with the requirements of UK legislation.

Advantages of good supply chain management

▶ reduction of waste of materials and time;
▶ ability to respond rapidly to changing requirements at short notice;
▶ reduction in accidents due to close liaison;
▶ legal duties addressed satisfactorily; and
▶ easier to maintain good customer care.

Role and functions of external agencies

Office of Rail Regulation (ORR)

▶ responsible for the independent health and safety regulation of the railway industry in the UK;
▶ inspectors known as Her Majesty's Railway Inspectors.

Fire and Rescue Authority

▶ enforces fire safety law;
▶ undertakes random fire inspections (often to examine fire risk assessments);
▶ can issue alteration, improvement and prohibition notices; and
▶ needs to be informed during the planning stage of building alterations when fire safety of the building may be affected.

Environment Agency (Scottish Environmental Agency)

▶ responsible for authorising and regulating emissions from industry;
▶ ensuring effective controls of the most polluting industries;
▶ monitoring radioactive releases from nuclear sites;
▶ ensuring that discharges to controlled waters are at acceptable levels;
▶ setting standards and issuing permits for the collection, transporting, processing and disposal of waste (including radioactive waste);

- ► enforcement of the Producer Responsibility Obligations (Packaging Waste) Regulations 1997; and
- ► enforcement of the Waste Electrical and Electronic Equipment (WEEE) Directive and its associated directives.

Insurance companies

- ► legal requirement for employers to insure against liability for injury or disease to their employees arising out of their employment;
- ► offer fire and public liability insurance; and
- ► can influence health and safety standards by weighting the premium offered to an organisation against its health and safety record.

Management of Health and Safety at Work Regulations

Employers' duties

- ► undertake suitable and sufficient written risk assessments (five or more employees);
- ► put in place effective health and safety management arrangements and record them if five or more employees;
- ► undertake preventative and protective measures on the basis of the principles of prevention specified in Schedule 1 to the Regulations;
- ► employ a competent health and safety person;
- ► develop suitable emergency arrangements and inform employees and others;
- ► provide health and safety information to employees and others, such as other employers, the self-employed and their employees who are sharing the same workplace, and parents of child employees or those on work experience;
- ► cooperate in health and safety matters with other employers who share the same workplace;
- ► provide employees with adequate and relevant health and safety training;
- ► provide temporary workers and their contract agency with appropriate health and safety information;

▶ protect new and expectant mothers and young persons from particular risks; and

▶ under certain circumstances, as outlined in Regulation 6, provide health surveillance for employees.

The information that should be supplied by employers under the Regulations is:

▶ risks identified by any risk assessments, including those notified to them by other employers sharing the same workplace;

▶ the preventative and protective measures that are in place; and

▶ the emergency arrangements and procedures and the names of those responsible for the implementation of the procedures.

Employees' duties

▶ use any equipment or substance in accordance with any training or instruction given by the employer;

▶ report to the employer any serious or imminent danger; and

▶ report any shortcomings in the employer's protective health and safety arrangements.

Self-employed duties

▶ carry out appropriate risk assessment;

▶ cooperate with other people who work in the premises and, where necessary, in the appointment of a health and safety coordinator; and

▶ provide comprehensible information to other people's employees working in their premises.

Joint occupation of premises

The Regulations specifically state that where two or more employers share a workplace – whether on a temporary or a permanent basis – each employer shall:

- cooperate with other employers;
- take reasonable steps to coordinate between other employers to comply with legal requirements; and
- take reasonable steps to inform other employers where there are risks to health and safety.

Responsibilities of clients and their contractors

Contractors are protected by the HSW Act (Section 3). Most contracted work is governed by a legally binding contract that should cover all parts of the work and the principles of cooperation, coordination and communication between organisations.

The following points should be considered when contractors are employed:

- Health and safety must be included in the contract specification.
- All significant hazards must be included in the contract specification.
- The contractor must be selected with safety in mind.
- Prior to the start of work, health and safety policies should be exchanged.
- The contractor must be given basic site and health and safety information, such as welfare and first-aid arrangements, significant hazards, safe storage of chemicals and the name of the contract supervisor.
- Where appropriate, the contractor should supply risk assessments and method statements.
- For construction work, the contractor should be aware of the position of buried/overhead services and the arrangements for the disposal of waste.
- The contractor should be monitored during the progress of the contract by the contract supervisor.
- The contract supervisor should check that the work has been completed safely at the end of the contract.

The management of construction activities (CDM 2015 Regulations)

The duty-holders and their main duties are:

- The **Client** has a major role and must ensure that:
 - all duty-holders are competent;
 - HSE is informed of notifiable projects;
 - pre-construction information is passed to relevant duty-holders (designers and contractors);
 - a Principal Designer is appointed and complies with their duties under the Regulations including the preparation of a health and safety file for the project;
 - a Principal Contractor is appointed and complies with their duties under the Regulations including the preparation of a construction phase plan;
 - the welfare requirements in Schedule 2 of the Regulations are followed;
 - health and safety arrangements are maintained and reviewed throughout the project; and
 - adequate resources and sufficient time are available for the safe completion of the work.
- **Domestic clients** are a special case and do not have specific duties under CDM 2015. Unless there is a written agreement between the domestic client and the principal designer, the duties of the client must be carried out by the contractor or principal contractor where there is more than one. Where no appointment is made, the first designer and contractor appointed are deemed to be the principal one in each case.
- The **Designer** must:
 - ensure that adequate health and safety provision is incorporated into the design
 - provide adequate information about any significant risks associated with the design; and
 - coordinate their work with that of the other duty-holders.
- The **Principal Designer** must ensure:
 - the project is carried out safely so far as is reasonably practicable;

▷ assistance is given to the client for the provision of pre-construction information;

▷ health and safety risks are identified, eliminated, controlled or reduced;

▷ the cooperation of everyone working on the project;

▷ the preparation and revision as necessary of the health and safety file;

▷ the provision of pre-construction information to every designer and contractor; and

▷ liaison with the principal contractor for the duration of the project particularly over the preparation of the construction phase plan.

▶ The **Principal Contractor** must ensure that:

▷ the health and safety construction phase plan is developed and managed;

▷ the construction phase is properly planned, managed and monitored;

▷ only competent sub-contractors are employed and supplied with necessary health and safety information;

▷ necessary site health and safety rules are drawn up and suitable site induction is provided;

▷ access by unauthorised persons is controlled preventing unauthorised access to the site;

▷ there is liaison with the principal designer throughout the project particularly regarding information for the health and safety file and management of the construction phase plan;

▷ the health and safety file is appropriately updated, reviewed and revised to take account of changes;

▷ suitable welfare facilities are provided on site; and

▷ the project notification is displayed.

▶ The **Contractor** must:

▷ not undertake construction work on a project unless satisfied that the client is aware of their duties under CDM 2015;

▷ plan, manage and monitor the way construction work is done to ensure it is safe so far as is reasonably practicable;

▷ where there is no principal contractor, ensure that a construction phase plan is drawn up prior to setting up the site;

▷ provide any employees or persons under their control, any information and instruction to ensure the safety of the project;

▷ ensure that the welfare requirements in Schedule 2 of the Regulations are provided for employees and others under their control;

▷ comply with any directions given by the principal designer or principal contractor and any site rules; and

▷ consult with their workers.

Required documentation

1 **Pre-construction information** – needs to identify the hazards and risks associated with the design and construction work.

2 **Health and safety file** – is a record of health and safety information required by the subsequent users of the finished construction project.

3 **Construction health and safety plan** – defines the organisation and arrangements required to control the site risks and coordinate the construction work.

4 **Notifiable work notice** – issued to the HSE if the construction work is to last longer than 30 days and has more than 20 workers simultaneously or involves more than 500 person days of work.

The principles of assessing and managing contractors

Scale of contractor use

The use of contractors has increased over recent years. Reasons include:

▶ the need to supplement permanent staff particularly for specialist tasks, or to undertake non-routine activities;

▶ the demand for products or services is uncertain;

30

▶ when more flexibility is required;
▶ contractors usually supply their own tools and equipment associated with the contract;
▶ there are no permanent staff available to perform the work;
▶ the financial overheads and legal employment obligations are lower;
▶ most of the costs associated with increasing and reducing employee numbers as product demand varies do not relate to contractors; and
▶ permanent staff can concentrate on the core business of the organisation.

There are some disadvantages in the use of independent contractors including:

▶ contractors/sub-contractors may cost more than the equivalent daily rate for employing a worker;
▶ by relying on contractors, the skills of permanent staff are not developed;
▶ there is less control over contractors than permanently employed staff. This can be a significant problem if the contractor sub-contracts some of the work; and
▶ the control of the contractor and the quality of the work is crucially dependent on the terms of the contract.

Selection of contractors

Most contracted work is governed by a legally binding contract and it is, therefore, very important that the contract covers all parts of the work – for example competent workers, safe access when working at height, fire precautions and safe waste disposal.

The following issues should be considered:

▶ an adequate health and safety policy;
▶ competent supervision;
▶ the availability of competent health and safety advice;
▶ past accident record;
▶ ability to assess hazards and risks involved in the contract and implement appropriate control measures;

31

▶ examples of method statements;
▶ a competent, trained and experienced workforce; and
▶ good references from previous contracts.

Management and authorisation of contractors

It is important that contractors are made aware of:

▶ the health and safety procedures and rules of the organisation;
▶ the hazards on your site, particularly those associated with the project;
▶ any special equipment or personal protective equipment that they need to use;

▶ the emergency procedures and the sound of the alarm; and
▶ the safe disposal of waste.

Contractors, their employees, sub-contractors and their employees, should not be allowed to commence work on any client's site without **authorisation** signed by the company contact. The contact will need to check as a minimum the following:

▶ the correct contractor for the work has been selected;
▶ the contractor has made appropriate arrangements for supervision of staff;
▶ the contractor has received and signed for a copy of the contractor's safety rules;
▶ the contractor is clear what is required, the limits of the work and any special precautions, including permits to work, that need to be taken; and
▶ the contractor's personnel are properly qualified for the work to be undertaken.

Copies of the authorisation document should be retained by all contractors and a copy kept on site.

Safety rules for contractors

Contractors' safety rules should contain as a minimum the following points:

▶ **Health and safety** – that the contractor operates to at least the minimum legal standard and conforms to accepted industry good practice.

▶ **Supervision** – that the contractor provides a good standard of supervision of their own employees.

▶ **Sub-contractors** – that they may not use sub-contractors without prior written agreement from the organisation.

▶ **Authorisation** – that each employee must carry an authorisation card issued by the organisation at all times while on site.

1.2

Health and safety management systems 1 – Plan

Learning outcomes

Outline the key elements of a health and safety management system ☐

Explain the purpose and importance of setting policy for health and safety ☐

Describe the key features and appropriate content of an effective health and safety policy ☐

🔑 Key revision points

The four key elements of the health and safety management system HSG65	☐
The legal requirements for a health and safety policy	☐
The three key elements of a health and safety policy	☐
Target setting for health and safety performance	☐
Circumstances leading to the need for a review of health and safety policy	☐
The reasons for unsuccessful health and safety policies	☐

Key elements of a health and safety management system (HSG65)

The HSE document HSG65, Managing for health and safety, describes the occupational health and safety management system used extensively in the UK and is based on a 'Plan, Do, Check, Act' approach.

The four elements of HSG65 are:

1 **PLAN** – the establishment of standards for health and safety management based on risk assessment and legal requirements and the formulation of a health and safety plan and policy.

2 **DO** – the implementation of the health and safety plan to achieve the policy objectives and standards. This will involve good communication, the promotion of competency, the commitment of all employees and a responsive reporting system.

3 **CHECK** – the measurement of progress with plans and compliance with standards and includes both active (sometimes called proactive) and reactive monitoring of the health and safety management system. It will measure the performance of the organisation against its own long-term goals and objectives and its progress with the health and safety plan and compliance with standards.

4 **ACT** – review against objectives and standards and take appropriate action. The results of monitoring and independent audits are used to indicate whether the objectives and targets set in the health and safety policy need to be changed. Changes in the health and safety environment in the organisation, such as an accident, should also trigger a performance review.

The framework of health and safety management HSG65

HSG65 element	Main topics in each element	Components of each topic
PLAN	Health and safety policy	Define health and safety aims, objectives and resources. Hazard identification and some risk assessment may be required.
	Health and safety plan	Plan should include emergency procedures (including first-aid), legal requirements and methods of performance measurement.
DO	Profile of health and safety risks	Assess risks. Introduce and/or monitor risk controls.
	Health and safety organisation	Designation of health and safety responsibilities.
		Ensure good communication, consultation, supervision and training of the workforce.
	Implementation of health and safety plan	Develop safe systems of work and, where appropriate, permit to work.
		Provide adequate resources, including competent advice where needed.
CHECK	Performance measurement	Monitor the implementation of the health and safety plan using various inspection techniques.
	Accident and incident investigation	Investigate the causes of accidents, incidents or near misses and monitor any recommendations made.
ACT	Performance review	Review health and safety performance using accidents and incidents, ill-health data and relative to benchmarks from similar organisations.
	Act on lessons learnt	Review plans, policy documents and risk assessments to see if they need updating.
	Continual improvement	Take any required action following audit and inspection reports.

Planning a health and safety management system

The health and safety planning process begins with finding:

▶ the correct information about the existing management system for health and safety;

▶ suitable benchmarks against which to make comparisons; and

▶ competent people to carry out the analysis and make sensible judgements.

Further judgement may be necessary to establish if the system is:

▶ adequate for the organisation and the range of hazards/risks;

▶ working as intended and achieving the right objectives; and

▶ delivering cost-effective and proportionate risk control in the workplace.

Purpose and importance of setting policy for health and safety

The purpose of setting a health and safety policy is to ensure that:

▶ everybody associated with the organisation is aware of its health and safety aims and objectives and how they are to be achieved;

▶ the performance of the organisation is enhanced in areas other than health and safety;

▶ there is effective personal development of the workforce;

▶ business efficiency is improved throughout the operation; and

▶ the involvement of senior management in health and safety issues is evident to all stakeholders.

The policy should state the intentions of the business in terms of clear aims, objectives, organisation, arrangements and targets for all health and safety issues.

The law requires that the written health and safety policy should include the following three sections:

▶ a health and safety **policy statement of intent** which includes the health and safety aims and objectives of the organisation;

▶ the health and safety **organisation** detailing the people with specific health and safety responsibilities and their duties; and

▶ the health and safety **arrangements** in place in terms of systems and procedures.

Setting health and safety objectives

The standard of the health and safety objectives of an organisation will depend upon:

▶ the seniority of the person setting the objectives;

▶ the formal documentation associated with the objectives and their relevance at each functional level in the organisation;

▶ the prioritisation of targets to define the Key Performance Indicators (KPIs);

▶ the incorporation of relevant legal, technological options and good practice guidance;

▶ the identification and assessment of all significant hazards and risks within the organisation;

▶ the success with which they integrate with financial, operational and business requirements; and

▶ the incorporation of the views of employees, stakeholders and other interested parties.

The objectives must be **specific**, **measurable**, **achievable** and with **realistic timescales** (**SMART**) and kept up to date with any changes in legislation.

Legal requirements

Section 2 of the HSW Act and the Employers' Health and Safety Policy Statements (Exception) Regulations 1975 require employers with **five or more** employees to prepare and review on a regular basis a written health and safety policy, together with the necessary organisation and arrangements to carry it out and to bring the policy and any revision of it to the notice of their employees.

Policy statement of intent	Organisation	Arrangements to include
▶ Aims and objectives	▶ Safety manual	▶ Planning and organising
▶ Duties of employer and employees	▶ Organisational chart	▶ Accident reporting
▶ Performance targets and benchmarks	▶ Responsibilities	▶ Emergencies
▶ Name of person responsible for health and safety	▶ Allocation of resources including finance	▶ Contractors and visitors
▶ Posted and dated	▶ Safety monitoring system	▶ Consultation and communication with employees
	▶ Identification of main hazards	▶ Fire precautions
		▶ Main risk assessments and hazard control
		▶ Performance monitoring

Reasons for a review of the health and safety policy

▶ Significant organisational and/or technological changes have taken place.

▶ There have been changes in personnel and/or legislation.

▶ Health and safety performance has fallen below the occupational group's benchmarks.

▶ The monitoring of risk assessments and/or accident/incident investigations indicates that the health and safety policy is no longer totally effective.

▶ Enforcement action has been taken by the HSE or Local Authority.

▶ A sufficient period of time has elapsed since the previous review.

Benchmarking

Benchmarking compares the performance of the organisation with that of similar organisations. The advantages of benchmarking are that:

▶ the key performance indicators for a particular organisation may be easily identified;
▶ it helps with continuous improvement;
▶ it focuses attention on weaker performance areas;
▶ it gives confidence to various stakeholders; and
▶ it is useful feedback for boards, chief executives and managers.

The effects of a positive health and safety performance

are to:

▶ support the overall development of personnel;
▶ improve communication and consultation throughout the organisation;
▶ minimise financial losses due to accidents and ill-health and other incidents;
▶ directly involve senior managers in all levels of the organisation;
▶ improve supervision, particularly for young persons and those on occupational training courses;
▶ improve production processes; and
▶ improve the public image of the organisation or company.

The reasons for unsuccessful health and safety policies

include:

▶ the statements in the policy and the health and safety priorities are not understood by, or properly communicated to, the workforce;
▶ minimal resources are made available for the implementation of the policy;
▶ too much emphasis on rules for employees and too little on management policy;

- a lack of parity with other activities of the organisation (such as finance);
- lack of senior management involvement in health and safety;
- inadequate personal protective equipment;
- unsafe and poorly maintained machinery and equipment; and
- a lack of health and safety monitoring procedures.

Health and safety management systems 2 – Do 1

Learning outcomes	
Outline the organisational health and safety roles and responsibilities of employers, directors and managers	☐
Explain the concept of health and safety culture and its significance in the management of health and safety in an organisation	☐
Outline the human factors which influence behaviour at work in a way that can affect health and safety	☐
Explain how health and safety behaviour at work can be improved	☐

🔑 Key revision points

The health and safety roles and responsibilities of employers to their employees and others affected by their undertaking ☐

The health and safety roles and responsibilities of directors, managers and supervisors ☐

The competence and responsibilities of the health and safety practitioner/adviser ☐

The definition and importance of a health and safety culture ☐

The relationship between culture and performance ☐

The definition of human factors and their influence on the culture ☐

The development of a positive health and safety culture ☐

The importance of good communication ☐

The duties of employers to consult with the workforce ☐

The different forms of health and safety training ☐

The internal and external influences on the health and safety culture of an organisation ☐

The duties of employers to consult with the workforce ☐

Organisational health and safety roles and responsibilities of employers, directors and managers

Organisational health and safety responsibilities – employers

The organisational health and safety responsibilities of employers are closely linked to their statutory duties which are covered in detail in 1.1.

Organisational health and safety responsibilities – directors

Directors and Board members should ensure that:

▶ health and safety arrangements are properly resourced;
▶ competent health and safety advice is obtained;
▶ regular reports are received on health and safety performance;
▶ any new or amended health and safety legislation is implemented;
▶ risk assessments are undertaken;
▶ there are regular audits of health and safety management systems and risk control measures; and
▶ there is adequate consultation with employees on health and safety issues.

The four elements that boards need to incorporate into their management of health and safety are:

▶ planning the direction of health and safety;
▶ delivering the plan for health and safety;
▶ monitoring health and safety performance; and
▶ reviewing health and safety performance.

Effective health and safety performance comes from the top, and directors have both collective and individual responsibility for health and safety.

Directors and Board members must ensure that	Management of health and safety at Board level involves
▶ the health and safety of employees and others, such as members of the public, is protected; ▶ risk management includes health and safety risks and becomes a key business risk in board decisions; and ▶ health and safety duties imposed by legislation are followed.	▶ planning the direction of health and safety; ▶ delivering the plan for health and safety; ▶ monitoring health and safety performance; and ▶ reviewing health and safety performance.

Organisational health and safety responsibilities – managers

Managing directors/chief executives, line managers and supervisors play key roles in ensuring that the health and safety policy is delivered and monitored.

1. Managing Directors/Chief Executives

Managing Directors/Chief Executives are responsible for:

▶ the health and safety performance within the organisation;
▶ ensuring that adequate resources are available for the health and safety requirements within the organisation, including the appointment of a senior member of the senior management with specific responsibility for health and safety;
▶ appointing one or more competent persons and adequate resources to provide assistance in meeting the organisation's health and safety obligations, including specialist help where appropriate;
▶ the establishment, implementation and maintenance of a health and safety programme for the organisation that encompasses all areas of significant health and safety risk;
▶ the approval, introduction and monitoring of all site health and safety policies, rules and procedures; and
▶ the review and possible revision annually of the effectiveness of the health and safety programme.

2. Departmental Managers

The principal departmental managers may report to the Site Manager, Managing Director or Chief Executive. In particular, they:

▶ are responsible and accountable for the health and safety performance of their department;

▶ are responsible for the engagement and management of contractors and that they are properly supervised;

▶ must ensure that any machinery, equipment or vehicles used within the department are maintained, correctly guarded and meet agreed health and safety standards; copies of records of all maintenance, statutory and insurance inspections must be kept by the Departmental Manager;

▶ develop a training plan that includes specific job instructions for new or transferred employees and follow up on the training by supervisors; copies of records of all training must be kept by the Departmental Manager;

▶ personally investigate all lost workday cases and dangerous occurrences and report to their line manager; progress any required corrective action.

3. Supervisors

The supervisors are responsible to and report to their Departmental Manager. In particular, they:

▶ are responsible and accountable for their team's health and safety performance;

▶ enforce all safe systems of work procedures that have been issued by the Departmental Manager;

▶ instruct employees in relevant health and safety rules, make records of this instruction and enforce all health and safety rules and procedures;

▶ supervise any contractors that are working within their area of supervision;

▶ enforce personal protective equipment requirements, check that it is being used and periodically appraise condition of equipment; and

47

▶ record any infringements of the personal protective equipment policy.

Health and safety adviser

must:

▶ be competent following the attainment of a health and safety qualification and training;
▶ report directly to senior management on matters of policy;
▶ keep up to date with technological advances and legislative changes;
▶ advise on the establishment of health and safety, maintenance and accident investigation procedures; and
▶ provide liaison with external agencies, such as the HSE, Fire Authorities, contractors, insurance companies and the public.

Concept of health and safety culture

The safety culture of an organisation is the product of individual and group values, attitudes, perceptions, competencies and patterns of behaviour that determine the commitment to, and the style and proficiency of, an organisation's health and safety management.

Features of a good health and safety culture

▶ Leadership and commitment to health and safety throughout the organisation, which is demonstrated in a genuine and visible way;
▶ mutual trust throughout the organisation;
▶ acceptance that high standards of health and safety are achievable as part of a long-term strategy formulated by the organisation requiring sustained effort and interest;
▶ detailed risk assessments and control and monitoring procedures;
▶ a health and safety policy statement that conveys a sense of optimism and outlines short- and long-term health and safety objectives; such a policy should also include codes of practice and required health and safety standards;

- training, communication and consultation procedures to ensure ownership and participation in health and safety throughout the organisation;
- encouragement to the workforce to report potential hazards;
- health and safety systems for monitoring equipment, processes, behaviour and procedures and the prompt rectification of any defects; and
- prompt accident investigation and implementation of remedial actions.

Indicators of a health and safety culture

- accident/incident rates;
- accident frequency rate;
- sickness and absentee rates;
- resources available for health and safety;
- level of legal and other compliance;
- turnover rates for employees;
- level of complaints;
- selection and management of contractors;
- levels and effectiveness of communication and supervision;
- health and safety management structure; and
- level of insurance premiums.

Human factors

1 in 10 near misses = 1 accident.

90% accidents due to human error – 70% poor management.

Human factors are affected by the:

- Organisation
- Job
- Individual (personal) factors.

Organisation

- must have a positive health and safety culture;
- manage health and safety by providing leadership and involvement of senior managers;

49

▶ motivate the workforce to improve health and safety performance; and

▶ measure health and safety performance.

Job

▶ recognise possibility of human error;

▶ good ergonomics, equipment design and layout of workstation;

▶ clear job descriptions;

▶ safe systems of work and operating procedures;

▶ job rotation and regular breaks;

▶ provision of correct tools; and

▶ effective training schedule and good communication.

Individual (personal) factors

The three common psychological factors are:

▶ Attitude – tendency to behave in a particular way in a given situation, influenced by social background and peer pressure

▶ Motivation – the driving force behind the way a person acts or is stimulated to act

▶ Perception – the way in which a person believes or understands information supplied or a situation.

Other related factors:

▶ self-interest – e.g. effect of bonus systems;

▶ position in the team;

▶ acknowledgement by management of good work/initiative;

▶ hearing and/or memory loss;

▶ experience and competence;

▶ age, personality, attitude, language problems;

▶ training undertaken and information given;

▶ effect of shift working – e.g. night working; and

▶ health (physical and mental).

Language and communication issues may be particular problems for some migrant workers.

The following **negative** factors can also affect the individual:

▶ low skill and competence levels;

▶ tired colleagues;

▶ bored or disinterested colleagues;

▶ individual medical or mental problems;

▶ complacency from repetitive tasks and lack of awareness training;

▶ inexperience especially if the employee is new or a young person; and

▶ peer pressure to conform to the 'group' or an individual's perception of how a task should be completed.

Human errors may be:	Violations may be:
1 Slips – failure to carry out the correct actions of a task	1 Routine – the breaking of a safety rule or procedure is the normal way of working
2 Lapses – failure to carry out particular actions that form part of a working procedure	2 Situational – job pressures at a particular time make rule compliance difficult
3 Mistakes are: ▶ rule-based – a rule or procedure is applied or remembered incorrectly or ▶ knowledge-based – well-tried methods or calculation rules are applied incorrectly	3 Exceptional – a safety rule is broken to perform a new task

Methods of improving health and safety behaviour at work

Development of a positive health and safety culture

By:

1 Commitment of management is the most important factor:

 ▶ Proactive management

 ▶ Promotion by example (e.g. wearing PPE)

2 The promotion of health and safety standards for:

 ▶ Selection and design of premises

51

- ▶ Selection and design of plant, processes and substances
- ▶ Recruitment of employees and contractors
- ▶ Risk assessments and control implementation
- ▶ Competence, maintenance and supervision
- ▶ Emergency planning and training
- ▶ Transportation of the product and its subsequent maintenance and servicing

3 Competence of the workforce, including in health and safety:

- ▶ Knowledge and understanding of the work/job
- ▶ Capacity to apply knowledge to the particular task
- ▶ Awareness of one's limitations.

Identifying and keeping up to date with legal requirements

The comparison with legal standards and requirements is an essential part of the setting of health and safety objectives and can be achieved by:

- ▶ Regular checking of the HSE website;
- ▶ Free subscription to HSE bulletins on the internet;
- ▶ Reading various health and safety periodicals;
- ▶ Reading journals or websites of trade associations; or
- ▶ Reading the text of legislation online.

Communication

- ▶ Verbal (by mouth) – conversations, telephone
- ▶ Written – memos, emails, meeting minutes, data sheets
- ▶ Graphic – safety signs, posters, charts.

Use of notice boards

The limitations of notice boards include:

- ▶ the information may not be read;
- ▶ the notice boards may not be accessible;
- ▶ the information may become outdated or defaced;
- ▶ some employees may not be able to read while others may not understand what they have read;

- there may be language barriers;
- the information is mixed in with other non-health and safety information; and
- there is no opportunity offered for feedback.

Alternative methods of communication include:

- memos, emails and company intranet;
- toolbox talks and team briefings;
- induction training and any further back-up training sessions;
- newsletters, bulletins and payslips;
- digital video media including DVDs;
- a staff handbook; and
- through safety committees, safety representatives and representatives of employee safety.

The approved poster entitled '*Health and Safety Law – what you should know*' must be displayed on a notice board or the approved leaflet distributed.

In addition to the health and safety poster, the following types of health and safety information could be displayed on a workplace notice board:

- a copy of the Employer's Liability Insurance Certificate;
- details of first-aid arrangements;
- emergency evacuation and fire procedures;
- minutes of the last health and safety committee meeting;
- details of health and safety targets and performance against them; and
- health and safety posters and campaign details.

Barriers to effective communication

- language and dialect
- acronyms and jargon
- various physical and mental disabilities
- attitudes and perception of workers and supervisors.

Types of accident propaganda

▶ Statistics
▶ Films, DVDs and posters
▶ Targets
▶ Records.

For safety propaganda to be effective, it must have:

▶ a simple understandable message
▶ a positive believable message
▶ an appealing format that will motivate the reader.

Consultation with employees

Difference between informing and consulting:

▶ 'informing' is a one-way process involving the provision of relevant information by management to workers; whereas
▶ 'consulting' is a two-way process where account is taken of the views of workers before any decision is taken.

The benefits of consultation include:

▶ an improved health and safety culture;
▶ motivation of staff;
▶ reduction in accidents and ill-health; and
▶ improved overall performance of the organisation.

Safety representatives from a recognised trade union have the following rights:

▶ to investigate accidents and dangerous occurrences;
▶ to investigate health and safety complaints;
▶ to undertake workplace inspections;
▶ to receive information from health and safety inspectors;
▶ to attend health and safety committee meetings;
▶ to have access to suitable facilities to perform their functions; and
▶ to be allowed time off with pay for health and safety training.

If two or more representatives request in writing for a health and safety committee to be set up then the employer must comply within 3 months.

Representatives of Employee Safety (ROES) were established under the Consultation with Employees Regulations.

The employer must **inform** and **consult** employees on health and safety matters.

ROES have the following functions:

▶ inform the employer of health and safety concerns of the workforce;
▶ inform the employer of potential hazards and dangerous occurrences in the workplace;
▶ inform the employer of any general matters that affect the health and safety of the workforce; and
▶ speak on behalf of the workforce to health and safety inspectors.

Employer must consult on:

▶ new processes or equipment or any changes in them;
▶ the appointment arrangements for a health and safety competent person;
▶ the results of any risk assessments;
▶ the arrangements for the management of health and safety training; and
▶ the introduction of new technologies.

Types of information the employer does NOT need to disclose – if it:

▶ violates a legal prohibition
▶ endangers national security
▶ relates to a specific individual
▶ could harm the company commercially
▶ was obtained from legal proceedings.

Safety committees

Terms of reference should include the following:

▶ the study of accident and notifiable disease statistics to enable reports to be made of recommended remedial actions;
▶ the examination of health and safety audits and statutory inspection reports;

▶ the consideration of reports from the external enforcement agency;

▶ the review of new legislation, Approved Codes of Practice and guidance and their effect on the organisation;

▶ the monitoring and review of all health and safety training and instruction activities in the organisation;

▶ the monitoring and review of health and safety publicity and communication throughout the organisation;

▶ the development of safe systems of work and safety procedures;

▶ reviewing risk assessments;

▶ considering reports from safety representatives; and

▶ the continuous monitoring of arrangements for health and safety and revising them whenever necessary.

Types of health and safety training

▶ Induction – at recruitment

▶ Job-specific

▶ Supervisory and management

▶ Specialist (e.g. first-aid)

▶ Refresher or reinforcement.

Internal influences on health and safety culture	External influences on health and safety culture
Management commitment	Expectations of society
Production/service demands	Legislation and enforcement
Communication	Insurance companies
Competence	Trade unions
Employee representation	State of the economy
	Commercial stakeholders

1.4

Health and safety management systems 2 – Do 2

Learning outcomes

Explain the principles and practice of risk assessment ☐

Explain the general principles of control and a general hierarchy of risk reduction measures ☐

Identify the key sources of health and safety information ☐

Explain what factors should be considered when developing and implementing a safe system of work for general activities ☐

Explain the role and function of a permit-to-work system ☐

Outline the need for emergency procedures and the arrangements for contacting emergency services ☐

Outline the requirements for, and effective provision of, first-aid in the workplace ☐

Key revision points

The legal requirements for a health and safety risk assessment	☐
The meaning of 'suitable and sufficient'	☐
The forms and objectives of risk assessment	☐
Details of accident categories and types of health risks	☐
The risk assessment process and its management	☐
Groups that require a special risk assessment	☐
The principles of prevention and the hierarchy of risk control	☐
The development and application of safe systems of work and permits to work for various activities, including those in confined spaces	☐
Hazards and controls for confined spaces and machinery maintenance work	☐
The development of emergency and first-aid procedures in the workplace	☐

Principles and practice of risk assessment

Legal requirement

Regulation 3 of the Management of Health and Safety at Work Regulations requires written suitable and sufficient risk assessment if there are five or more employees.

Suitable and sufficient means:

▶ identify significant risks only;
▶ identify measures required to comply with legislation; and
▶ the assessment should remain appropriate and valid over a reasonable period of time.

Hazard – the potential to cause harm
Risk – the likelihood to cause harm
Residual risk – the risk remaining after some controls are in place.

Forms of risk assessment

▶ Quantitative – calculated from risk = likelihood × severity
▶ Qualitative – descriptor (high, medium or low) used to describe timetable for remedial action
▶ Generic – covers similar activities or work equipment.

Health risks

▶ Chemical – exhaust fumes, paint solvents, asbestos
▶ Biological – legionella, pathogens, hepatitis
▶ Physical – noise, vibration, radiation
▶ Psychological – stress, violence
▶ (Ergonomic – musculoskeletal disorders).

Risk assessment team requirements

▶ all need training in risk assessment
▶ leader should have health and safety experience
▶ all need to be competent to assess risks in area under examination
▶ all need to know their own limitations
▶ include local line manager in the team
▶ at least one team member with report writing skills.

Risk assessment process

▶ **Hazard identification – Step 1** of HSE's five steps.
▶ **Persons at risk – Step 2** of HSE's five steps:
 ▷ employees, agency/temporary workers, contractors, shift workers
 ▷ members of the public – visitors, customers, patients, students, children, elderly
 ▷ special groups – young persons, expectant or nursing mothers, workers with a disability, lone workers.

59

▶ **Evaluation of risk level (residual risk) – Step 3** of HSE's five steps:

 ▷ high, medium and low (defined qualitatively or quantitatively)
 ▷ both occupational and organisational risk levels need to be considered.

▶ **Detail risk controls (existing and additional) – Step 3** of HSE's five steps:

 ▷ the prioritisation of risk control is important
 ▷ risks can be reduced at the design stage by using the principles of prevention
 ▷ risks can be controlled by using the hierarchy of risk control.

Hierarchy of risk control

The health and safety management system ISO 45001 hierarchy of risk control is:

a eliminate the hazard;
b substitute with less hazardous materials, processes, operations or equipment;
c use engineering controls;
d use safety signs, markings and warning devices and administrative controls;
e use personal protective equipment.

There are several other similar hierarchies of risk control, which have been used over many years. A typical example is as follows:

1 elimination
2 substitution
3 changing work methods/patterns
4 reduced time exposure
5 engineering controls (isolation, insulation and ventilation)
6 good housekeeping
7 safe systems of work
8 training and information
9 personal protective equipment
10 welfare
11 monitoring and supervision
12 review.

Safety signs

- ▶ Red – prohibition – round (e.g. no smoking)
- ▶ Yellow – warning – triangular (e.g. wet floor)
- ▶ Blue – mandatory – round (e.g. ear defenders must be worn)
- ▶ Green – safe condition – square or rectangular (e.g. first-aid).
- ▶ **Record of risk assessment findings – Step 4** of HSE's five steps.
- ▶ **Monitor and review – Step 5** of HSE's five steps:
 Regular reviews required but need to be more frequent if:
 - ▷ new legislation introduced
 - ▷ new information available on substances or process
 - ▷ changes to the workforce – introduction of trainees
 - ▷ an accident has occurred.

Special cases

1. Young persons

- ▶ are under 18 years
- ▶ covered by Regulation 19 MHSWR
- ▶ subject to peer pressure and are inexperienced
- ▶ are eager to please
- ▶ appropriate level and approach of subject matter in training sessions.

There are additional requirements if under school leaving age.

A special risk assessment must be made to include details of:

- ▶ the work activity;
- ▶ any prohibited processes or equipment;
- ▶ the health and safety training provided; and
- ▶ the supervision arrangements.

2. Expectant and nursing mothers

There are restrictions on the type of work that can be undertaken.

Risks include:

- ▶ manual handling;
- ▶ chemical and biological agents;

61

- ionising radiation;
- passive smoking;
- lack of rest room facilities;
- temperature variations;
- prolonged standing or sitting;
- whole-body vibration;
- issues associated with the use and wearing of personal protective equipment;
- working excessive hours;
- night working; and
- stress and violence to staff.

3. Workers with a disability

- emergency arrangements including 'raising the alarm'; and
- adequate wheelchair access to fire exit.

The special risk assessment should identify:

- the jobs with particular health and fitness requirements;
- the types of disability that would make certain jobs unsuitable; and
- the staff whose disabilities would exclude them from undertaking those jobs safely.

4. Lone workers

- special risk assessment (including violence);
- must be fit to work alone;
- special training should be given;
- possible to handle all equipment and substances alone;
- periodic visits by supervisor;
- regular mobile phone contact with base;
- first-aid arrangements; and
- emergency arrangements.

Sources of information on health and safety

Internal sources	External sources
▶ accident and ill-health records and investigation reports	▶ health and safety legislation
▶ absentee records	▶ HSE publications, such as Approved Codes of Practice, guidance documents, leaflets, journals, books and their website
▶ inspection and audit reports undertaken by the organisation and by external organisations such as the HSE	▶ international (e.g. ILO), European and British standards
▶ maintenance, risk assessment (including COSHH) and training records	▶ health and safety magazines and journals
▶ documents which provide information to workers	▶ information published by trade associations, employer organisations and trade unions
▶ any equipment examination or test reports	▶ specialist technical and legal publications
	▶ information and data from manufacturers and suppliers
	▶ the internet and encyclopedias

General principles of control

The planning and implementing section of the health and safety management system is based on risk assessment and concerns all the actions taken to control or eliminate hazards and reduce risks. The principles of prevention are used when equipment or processes are being designed or selected. The hierarchy of risk control enables risk to be further controlled.

The principles of prevention

▶ avoid risks
▶ evaluate risks which cannot be avoided
▶ adapt work to the individual
▶ adapt to technical changes
▶ replace dangerous items with less dangerous items
▶ develop an overall prevention policy
▶ give priority to collective measures (Safe Place strategy)
▶ give instructions to employees (Safe Person strategy).

Development and implementation of safe systems of work

Employer's duty

This is contained in the common law duty of care and in Section 2 of the Health and Safety at Work etc. Act.

Role of competent persons

A competent person or safety adviser should:

▶ assist managers to draw up guidelines for safe systems of work;
▶ prepare suitable documentation; and
▶ advise management on the adequacy of the safe systems produced.

Role of managers

▶ provide safe systems of work;
▶ ensure that employees are adequately trained in a specific safe system of work and are competent to carry out the work safely; and
▶ provide sufficient supervision to ensure that the system of work is followed and the work is carried out safely.

Employee involvement

Includes:

▶ consultation with those employees who will be exposed to the risks, either directly or through their representatives;
▶ discussion of the proposed system with those who will have to work under it and supervise it; and
▶ understanding that employees have a responsibility to follow the safe system of work.

Steps in the development of a safe system of work

1 Assess the task (complexity, accident records, etc.)
2 Identify the significant hazards associated with the task
3 Define safe methods for performing the task (including emergency procedures) – document the methods if required
4 Implement the safe system of work (written safe system to be signed off)
5 Monitor the safe system of work and review it if necessary
6 Train the workforce in safe procedures and enter on training record.

Safe systems of work are particularly important for:

▶ maintenance work
▶ contractors
▶ emergency procedures
▶ lone working
▶ vehicle operations
▶ cleaning operations.

Method statements are formal written safe systems of work and are often used in construction work.

The safe system of work should be based on a thorough analysis of the job or operation to be covered by the system. After the introduction of a safe system of work the following controls are required:

▶ Engineering or process controls (e.g. guarding)
▶ Documented procedures
▶ Behavioural controls requiring a certain standard of behaviour from individuals.

Procedures for lone workers may include:

▶ periodic visits from the supervisor;
▶ regular voice contact;
▶ automatic warning devices to alert others of problems;
▶ checks that the lone worker has returned safely;
▶ special arrangements for first-aid to deal with minor injuries; and
▶ emergency arrangements.

Specific examples of safe systems of work

Confined spaces

A confined space is defined in **Confined Spaces Regulations** as:

> '*any place, including any chamber, tank, vat, silo, pit, trench, pipe, sewer, flue, well or other similar space in which, by virtue of its enclosed nature, there arises a reasonably foreseeable specified risk.*'

The principal hazards associated with a confined space include:

▶ difficult access and egress;
▶ asphyxiation due to oxygen depletion;
▶ poisoning by toxic substance or fumes;
▶ explosions due to gases, vapours and dust;
▶ fire due to flammable liquids;
▶ fall of materials leading to possible head injuries;
▶ free-flowing solid such as grain in a silo;
▶ electrocution from unsuitable equipment;
▶ difficulties of rescuing injured personnel;
▶ drowning due to flooding; and
▶ fumes from plant or processes entering confined spaces.

The following topics need to be addressed in a safe system of work for a confined space:

▶ the appointment of a supervisor;
▶ the competence and experience required of the workers in the confined space;
▶ the isolation of mechanical and electrical equipment in an emergency;
▶ the size of the entrance must enable rapid access and exit in an emergency;
▶ the provision of adequate ventilation;
▶ the testing of the air inside the space to ensure that it is fit to breathe; if the air inside the space is not fit to breathe, then breathing apparatus will be essential;

▶ the provision of special tools and lighting, such as extra-low-voltage equipment, non-sparking tools and specially protected lighting;

▶ the emergency arrangements to cover the necessary equipment, training, practice drills and the raising of the alarm; and

▶ adequate communications arrangements to enable communication between people inside and outside the confined space and to summon help in an emergency.

Lone working

Typical control procedures may include:

▶ documented records of the location or itineraries of the lone workers;

▶ periodic visits from the supervisor to observe what is happening;

▶ regular voice contact, using mobile phones or radios, between the lone worker and the supervisor;

▶ automatic warning devices to alert others if a specific signal is not received from the lone worker;

▶ other devices to raise the alarm, which are activated by the absence of some specific action;

▶ checks that the lone worker has returned safely home or to their base;

▶ special arrangements for first-aid to deal with minor injuries – this may include mobile first-aid kits;

▶ arrangements for emergencies – these should be established and employees trained.

Role and function of a permit to work

Permits to work are:

▶ **formal** safe systems of work;

▶ required to be signed on/off by a responsible person; and

▶ often require equipment to be locked on/off by a responsible person.

To be used whenever there is a high risk of serious injury, such as:

▶ confined spaces
▶ live electricity (particularly high voltage)
▶ hot working (e.g. welding work)
▶ some machinery maintenance work.

Typical responsible persons are:

▶ site manager
▶ senior authorised person – often the chief engineer
▶ authorised person – issues permit
▶ competent person – receives permit
▶ operatives – supervised by a competent person
▶ specialists (e.g. electrical engineer)
▶ engineers – usually responsible for the work
▶ contractors.

The permit document should specify the following key items of information:

▶ the date, time and duration of the permit;
▶ a description and assessment of the task to be performed and its location;
▶ the plant/equipment involved, and how it is identified;
▶ the authorised persons to do the work;
▶ the steps which have already been taken to make the plant safe;
▶ potential hazards which remain, or which may arise as the work proceeds;
▶ the precautions to be taken against these hazards;
▶ the length of time that the permit is valid;
▶ the equipment to be released to those who are to carry out the work.

The permit will also include spaces for:

▶ signature certifying that the work is complete; and
▶ signature confirming re-acceptance of the plant/equipment.

Typical work tasks that might require a permit to work

Hot work

Typical controls include;

▶ a suitable fire extinguisher nearby;
▶ prompt removal of flammable waste material; and
▶ the damping down of nearby wooden structures such as floors.

Work on high-voltage apparatus (including testing)

Work on high-voltage apparatus (over about 600 V) is potentially high risk.

Hazards include:

▶ possible fatal electric shock/burns to the people doing the work;
▶ electrical fires/explosions;
▶ consequential danger from disruption of power supply to safety-critical plant and equipment.

Confined spaces

Examples include underground chamber, silo, trench, sewer, tunnel.

Hazards are:

▶ lack of oxygen and asphyxiation
▶ poor ventilation
▶ presence of fumes
▶ poor means of access and escape
▶ drowning
▶ claustrophobia
▶ electrical equipment (needs to be flameproof)
▶ presence of dust (e.g. silos)
▶ heat and high temperatures
▶ fire and/or explosion
▶ poor or artificial lighting.

69

Controls include:

▶ permit to work
▶ risk assessment
▶ training and information for all workers entering the confined space
▶ emergency arrangements in place
▶ emergency training
▶ no entry for unauthorised persons.

Machinery maintenance work

Hazards are:

▶ no perceived risk
▶ no safe system of work
▶ poor communications
▶ failure to brief contractors
▶ lack of familiarity
▶ poor design.

Control of hazards requires:

▶ effective planning
▶ a written safe system of work/permit to work
▶ a risk assessment to assess, control and reduce risks
▶ monitoring to ensure that the system of work and controls are used
▶ an effective training programme for all involved in the work.

Work at height

A permit to work may be required for some hazardous work at height such as roof work to ensure that a fall arrest strategy is in place. This is particularly important where:

▶ there are no permanent work platforms with fixed handrails on flat roofs;
▶ there are sloping or fragile roofs;
▶ specialist access equipment, like rope hung cradles are required; and
▶ access is difficult.

Emergency procedures

Examples of types of emergencies

- fire
- explosions, bomb scares
- escape of toxic gases
- major accident.

Typical elements of emergency procedures

- Fire notices and fire procedures (including testing)
- Fire drills and evacuation procedures
- Assembly and roll-call
- Arrangements for contacting emergency and rescue services
- Provision of information for emergency services
- Internal emergency organisation – including control of spillages and clean-up arrangements
- Media and publicity arrangements
- Business continuity arrangements.

Testing and training for emergencies

Training courses should be given to staff on a regular basis on the following topics:

- use of fire-fighting equipment;
- regular refresher first-aid training;
- suitable training (and competency assessment) for all those allocated particular roles in an emergency; and
- regular timed fire drills.

Provision of first-aid

Main functions of first-aid treatment

- preservation of life and/or minimisation of the consequences of serious injury until medical help is available
- treatment of minor injuries not needing medical attention.

Main first-aid requirements

▶ qualified first-aiders
▶ adequate facilities and equipment to administer first-aid
▶ an assessment of required first-aid cover and requirements
▶ an appointed person available to assist first-aiders.

Basic first-aid provision (including number of first-aiders)

Depends on:

▶ number of workers;
▶ the hazards and risks in the workplace;
▶ accident record and types of injuries;
▶ proximity to emergency medical services; and
▶ working patterns (shift work).

First-aider requires first-aid training and the first-aid course depends on the level of risk in the workplace.

▶ For low risk, a 6-hour Emergency First-aid at Work course
▶ For higher risk, 18-hour First-aid at Work course
▶ Either 6-hour emergency or 4-day first-aid course initially
▶ A repeat course every 3 years.

Appointed person has some other first-aid experience/qualification.

First-aid box – contents depend on particular workplace needs but should be checked regularly.

Health and safety management systems 3 – Check

Learning outcomes	
Outline the principles, purpose and role of active and reactive monitoring	☐
Explain the purpose of, and procedures for, investigating incidents (accidents, cases of work-related ill-health and other occurrences)	☐
Describe the legal and organisational requirements for recording and reporting incidents	☐

🔑 Key revision points

The reasons for measuring and monitoring health and safety performance	☐
The role of standards in the monitoring process	☐
The differences between reactive and active (proactive) monitoring	☐
The importance of clear, unambiguous report writing	☐
The reasons and legal requirements for recording and reporting incidents and accidents	☐
Basic accident investigation procedures and the different types of accident and incident	☐
Immediate and underlying causes of incidents	☐
Issues concerning insurance and compensation claims	☐

Active and reactive monitoring

Reactive monitoring (taking action after a problem occurs) involves	Proactive (or active) monitoring (taking action before problems occur) involves
▶ review of accident and ill-health reports – often to check that remedial advice has been actioned or ascertain trends and hot spots ▶ review of procedures following dangerous occurrences, other property damage and near misses ▶ review of compensation claims ▶ review of complaints from the workforce and members of the public ▶ review of procedures following enforcement reports and notices ▶ review of risk assessments following the discovery of additional hazards	▶ the active monitoring of the workplace for unsafe conditions ▶ the direct observation of workers for unsafe acts ▶ meeting with management and workers to discover any problems ▶ checking documents, such as maintenance records, near miss reports, insurance reports ▶ undertaking workplace inspections, sampling, surveys, tours and audits

Workplace inspections

Detailed check, often using a checklist, of the whole workplace and should cover:

▶ the premises (e.g. fire precautions, access/egress, housekeeping);

▶ the plant, equipment and substances (e.g. machine guarding, tools, ventilation);

▶ the procedures in place (e.g. safe systems of work, risk assessments, use of PPE); and

▶ the workforce (e.g. training, information, supervision, health surveillance).

Other issues with inspections:

▶ the competence of the observers;

▶ the frequency of inspections;

▶ the response to the inspection report; and

▶ the use of objective inspection standards.

Safety sampling – checking for safety defects in a selected area of the workplace (e.g. all fire extinguishers).

Safety surveys – a detailed inspection of a particular workplace activity throughout an organisation (e.g. manual handling).

Safety tours – an unscheduled brief inspection of a work area in the workplace by a team led by a senior manager.

Investigating incidents

Purpose of accident/incident investigation is to:

▶ eliminate the cause and future occurrences;

▶ determine the direct and indirect causes of the accident/incident;

▶ define any corrective and/or preventative actions;

▶ identify any deficiencies in risk controls, the health and safety management system and/or procedures;

▶ ensure that all legal requirements are being met; and

▶ comply with the recommendations of the Woolf Report so that essential information is available in the event of a civil claim.

Benefits of accident/incident investigation:

- prevention of a recurrence;
- prevention of future business losses;
- prevention of future increased insurance premiums and costs of criminal and civil actions; and
- improve employee morale and organisation reputation.

Accident triangles and their limitations

Accident triangles indicate the ratio between fatalities, serious and minor accidents (sometimes near misses are included). The limitations of these triangles are:

- similar injuries may have different causes;
- some incidents have the potential for serious injury yet may only result in a minor one;
- serious injuries have different underlying causes to minor injuries; and
- many injuries are musculoskeletal sprains and strains which do not result in fatal injuries.

Accident causation

Direct or immediate include	Indirect, root or underlying include
▶ unsafe acts by individuals due to poor behaviour or a lack of training, supervision, information or competence or a failure to wear PPE ▶ unsafe conditions, such as inadequate guarding, inadequate procedures, hazardous substances, ergonomic and/or environmental factors ▶ fire or explosive hazards	▶ poor machine maintenance and/or start-up procedures ▶ management and social pressures ▶ financial restrictions ▶ lack of management commitment to health and safety ▶ poor or lack of health and safety policy and standards ▶ poor workplace health and safety culture ▶ workplace and trade customs and attitudes

Elements of an investigation

▶ interview relevant operatives, managers, supervisors
▶ obtain detailed plans and/or photographs of scene
▶ check all relevant records (working procedures, maintenance, training, risk assessments)
▶ interview all witnesses
▶ possibly arrange for equipment and/or substances to be independently tested
▶ produce a concise report for management that includes
 ▷ background to the accident
 ▷ possible causes of the accident (direct and indirect)
 ▷ relevant health and safety legislation, guidance and standards
 ▷ recommendations (including any remedial actions)
 ▷ any additional training or follow-up requirements
▶ undertake a post-accident risk assessment.

Recording and reporting incidents

Accident reporting requirements

▶ the organisational accident/incident report form
▶ accident book
▶ RIDDOR 2013 – records kept for 3 years.

RIDDOR 2013

The Regulations apply to employees, the self-employed, trainees and contractors. Any fatality or injury to a member of the public resulting in off-site medical treatment is reportable.

Records must be kept of:

▶ any **accident, occupational disease or dangerous occurrence** which requires reporting under RIDDOR; and
▶ any other occupational accident causing injuries that result in a worker being away from work or incapacitated for more than three consecutive days. Over-three-day injuries do not have to be reported, unless the incapacitation period goes on to exceed seven days.

Fatality	▶ reportable by quickest means (telephone)
	▶ submit online form immediately
	▶ submit a report within 10 days
	▶ reportable if death occurs within 1 year of accident
Specified serious injury/ill-health	▶ reportable as soon as possible
	▶ submit online form immediately
	▶ submit a report within 10 days
	▶ examples – any fracture, other than fingers, thumbs or toes; any amputation; any other injury requiring admittance to hospital for more than 24 hours
Over 7 days' lost time injury	▶ submit online form within 15 days
Disease	▶ submit online form as soon as diagnosis is confirmed
	▶ examples – hepatitis; occupational dermatitis; hand–arm vibration syndrome
Dangerous occurrence	▶ a 'near miss' that could lead to serious injury or loss of life.
	▶ reportable by quickest means (telephone)
	▶ submit online form immediately
	▶ submit a report within 10 days
	▶ examples – collapse of scaffolding; the collapse, overturning or failure of any load-bearing part of lifts and lifting equipment; explosion or fire resulting in the suspension of normal working for more than 24 hours

Compensation and insurance issues

▶ importance of documentation and records;
▶ need for pre- and post-accident risk assessments;
▶ impact of Woolf Report and documentation required to defend a civil claim.

Health and safety management systems 4 – Act

Learning outcomes	
Explain the purpose of, and procedures for, health and safety auditing	☐
Explain the purpose of, and procedures for, regular reviews of health and safety performance	☐

🔑 **Key revision points**	
The role of performance review and audit	☐
The meaning, scope and purpose of health and safety audit	☐
The advantages and disadvantages of internal and external audit	☐
The reasons for measuring and monitoring health and safety performance	☐
The role of standards in the monitoring process	☐
The role of boards in performance review	☐
The measures to continually improve health and safety performance	☐

The final ACT steps in the health and safety management control cycle are auditing, performance review and continual improvement.

Health and safety auditing

Auditing – the independent collection of information on the efficiency, effectiveness and reliability of the whole health and safety management system measured against specific standards. It will check that the following are in place:

▶ appropriate management arrangements;
▶ adequate risk control systems exist and are implemented;
▶ appropriate workplace precautions; and
▶ appropriate documentation and records.

Audits should take place at regular intervals (every 2–4 years).

Other issues with audits

External audits:

- ▶ are independent of organisation;
- ▶ are competent;
- ▶ are familiar with external benchmarks;
- ▶ will not be inhibited from criticising members of management or the workforce;
- ▶ are more likely to be up to date with legal requirements and best practice in other companies;
- ▶ may be unfamiliar with the industry and seek unrealistic standards;
- ▶ are unlikely to be familiar with the workplace, tasks and processes;
- ▶ are more expensive than internal audits; and
- ▶ can offer bland reports.

Internal audits:

- ▶ are not independent of organisation and partiality may be questioned;
- ▶ may be subject to pressure from management and the workforce;
- ▶ usually require audit training;
- ▶ are less expensive than external audits;
- ▶ know the organisation well, particularly critical areas;
- ▶ are likely to be aware of what is practicable for the industry;
- ▶ have the ability to see improvements or a deterioration from the last audit;
- ▶ are familiar with the workforce and an individual's qualities and attitude;
- ▶ can spread good practice around the organisation; and
- ▶ may be unaware of external benchmarks.

Review of health and safety performance

Performance review is the final stage of the management process and reviews:

- ▶ all monitoring, inspection and audit reports;
- ▶ the adequacy of the health and safety management system, and performance standards against external benchmarks;
- ▶ whether new legislation or guidance has been applied;
- ▶ whether the health and safety policy objectives have been met or need modification to ensure continuous improvement; and
- ▶ whether there has been adequate feedback to/from managers.

Reviews should aim to include:

- ▶ evaluation of compliance with legal and organisational requirements;
- ▶ incident data, recommendations and action plans from investigations;
- ▶ inspections, surveys, tours and sampling;
- ▶ absences and sickness records and their analysis;
- ▶ any reports on quality assurance or environmental protection;
- ▶ audit results and implementation;
- ▶ monitoring of data, reports and records;
- ▶ communications from enforcing authorities and insurers;
- ▶ any developments in legal requirements or best practice within the industry;
- ▶ changed circumstances or processes;
- ▶ benchmarking with other similar organisations;
- ▶ complaints from neighbours, customers and the public;
- ▶ effectiveness of consultation and internal communications;
- ▶ whether health and safety objectives have been met; and
- ▶ whether actions from previous reviews have been completed.

The Board should review health and safety performance at least once a year. The review process should:

- ▶ examine whether the health and safety policy requires revision;
- ▶ examine whether risk management and other health and safety issues have been effectively reported to the Board;
- ▶ report health and safety shortcomings;

▶ address any weaknesses with effective measures and develop a system to monitor their implementation; and

▶ undertake immediate reviews in the light of major shortcomings or events.

Continual improvement

Continual improvement of safety performance will be achieved through:

▶ active and reactive evaluations of facilities, equipment, documentation and procedures through safety audits and surveys;

▶ active evaluation of each individual's performance to verify the fulfilment of their safety responsibilities; and

▶ a reactive evaluation in order to verify the effectiveness of the system for control and mitigation of risk.

Measures that could be used to improve safety management include:

▶ more succinct procedures;

▶ improved safety reviews, studies and audits;

▶ improved reporting and analysis tools;

▶ improved hazards identification and risk assessment processes and improved awareness of risks in the organisation;

▶ improved relations with the sub-contractors, suppliers and customers regarding safety;

▶ improved communication processes, including feedback from the personnel.

Managing and controlling hazards in construction activities

2.1

Construction law and management

Learning outcomes

Identify the scope, definition and particular issues relating to construction activities ☐

Outline the legal, moral and financial consequences of failing to manage health and safety within the construction industry ☐

Outline the scope and application of the Construction (Design and Management) Regulations 2015 ☐

Identify the nature and main sources of external construction health and safety information ☐

> **Key revision points**
>
> The scope of construction and the associated hazards and controls ☐
>
> The health and safety management of construction activities and the legal duties under the Construction (Design and Management) Regulations 2015 ☐
>
> Sources of external construction health and safety information ☐

The scope, definition and particular issues relating to construction activities

The scope of the construction industry includes the following types of work:

▶ general building work – domestic, commercial or industrial and may be:
 ▷ new building work, such as a building extension, or
 ▷ the refurbishment, alteration, maintenance or repair of existing buildings, or
 ▷ decommissioning, dismantling and demolition
▶ civil engineering projects involving road and bridge building, water supply and sewage schemes and river and canal work; and
▶ the use of woodworking workshops together with woodworking machines and their associated hazards.

It also covers the following range of activities:

▶ site clearance;
▶ loading, unloading and storage of materials;
▶ electrical installation and plumbing work;
▶ removal and maintenance of services (electricity, water, gas);
▶ landscaping;
▶ painting and decorating; and
▶ work in confined spaces such as excavations and underground chambers.

Particular issues related to construction:

▶ temporary nature of construction;
▶ transitory nature of the workforce;
▶ constantly changing nature of the construction site;
▶ employment of migrant workers;
▶ literacy and numeracy levels of workers;
▶ time pressures; and
▶ weather conditions.

Moral, legal and financial cases for health and safety management in the construction industry

Moral: high fatality rates, high accident rates and disease.
Legal: high number of enforcement notices, civil actions and prosecutions.
Financial: high direct and indirect costs.

Priorities and targets of the regulator:

▶ competence and control of contractors and employees;
▶ risk assessments, particularly for work at height;
▶ communication of risk controls to the workforce;
▶ selection, use, inspection and maintenance of equipment; and
▶ welfare facilities and site tidiness.

Scope and application of the Construction (Design and Management) Regulations 2015

In Element 1.1, the following topics were covered:

▶ The duties of the client, principal designer and the designers, principal contractor and the contractors, workers and domestic clients.
▶ The criteria for the notification of a construction project to the HSE.
▶ The meaning and contents of:
 ▷ pre-construction health and safety information;
 ▷ the construction phase health and safety plan; and
 ▷ the health and safety file.

89

However, the role of the principal designer will be covered in a little more detail since the role is new to CDM.

Principal designer

– is the designer with control over the pre-construction phase of the project and is appointed in writing by the client. The principal designer is usually an organisation (or on smaller projects an individual) that has:

a technical knowledge of the construction industry relevant to the project;

b the understanding and skills to manage and coordinate the pre-construction phase, including any design work carried out after construction begins; and

c where the principal designer is an organisation, the organisational capability to undertake the role and the necessary skills, knowledge and experience required of an individual designer.

The role of the principal designer involves:

▶ coordinating the work of others in the project team to ensure that significant and foreseeable risks are managed throughout the design process;

▶ liaison with the principal contractor for the duration of their appointment;

▶ during the pre-construction phase, sharing information that may affect the planning, management, monitoring and coordination of the construction phase – in particular, the information needed by the principal contractor to prepare the construction phase plan;

▶ extending this liaison into the construction phase to deal with ongoing design and to obtain information for the health and safety file; and

▶ arranging for the handover of the health and safety file to the principal contractor and alerting them to any issues that need to be reviewed and/or updated.

Other important aspects of CDM 2015

The appointment and required competence of relevant parties

involves the following:

1 All duty-holders must have the skills, knowledge and experience to undertake the work that secures health and safety. If an organisation is to be appointed, it must also have the appropriate organisational capability.

2 Clients must appoint principal designers and principal contractors as soon as practicable before the start of the construction phase, so they have enough time to plan and manage both the pre-construction and construction phases.

3 If a client fails to appoint either the principal designer or principal contractor, then the client must carry out their duties.

4 The required competence of the relevant parties is split into its component parts of skills, knowledge, training and experience, and – if it relates to an organisation – organisational capability.

5 The principal designer or contractor must have sufficient experience and a good track record in managing the health and safety risks involved in the project.

6 All individuals interviewed must have sufficient skills, knowledge and experience required for the post and have kept those capabilities up to date through continuing professional development.

Notification of projects

As covered in Element 1.1, a project is notifiable if the construction work on a construction site is scheduled to:

a last longer than 30 working days and have more than 20 workers working simultaneously at any point in the project; or

b exceed 500 person days.

The following additional points are important:

▶ Where a project is notifiable, the client must give notice in writing to the relevant enforcing authority as soon as is practicable before the construction phase begins.

▶ If a construction project is not notifiable at first, but there are subsequent changes to its scope so that it becomes notifiable, the client must notify the work to the relevant enforcing authority as soon as possible.

▶ Any modifications or updates to the notification should be sent to the enforcing authority making it clear that they relate to an earlier notification.

▶ The client must ensure that an up-to-date copy of the notice is displayed in the construction site office so that it is accessible to anyone working on the site.

▶ Notification may be submitted electronically.

Preparation of pre-construction information

1 The client has the main duty for the provision of pre-construction information. The principal designer must help the client in providing the pre-construction information to each designer and contractor appointed.

2 The client and the principal designer must also decide when designers and contractors will need pre-construction information to enable them to carry out their duties.

The construction phase plan

The client must ensure that:

▶ a construction phase plan for the project is prepared before the construction phase begins;

▶ the principal contractor (or, where relevant, the contractor) is provided with all the available relevant information, such as the pre-construction information, that they need to draw up the plan;

▶ the plan adequately addresses the arrangements for managing the risks; and

▶ the principal contractor (or contractor) regularly reviews and revises the plan to ensure that the plan takes account of any changes that occur as construction progresses and that it continues to be fit for purpose.

Before the start of the construction phase, the principal designer should regularly check that the principal contractor has the necessary information to prepare the plan. It must cover:

a the risks likely to arise during construction work;

b the measures needed to protect those affected by planning to provide and maintain:
 ▷ the right plant and equipment;
 ▷ the necessary information, instruction and training; and
 ▷ the right level of supervision;

c the resources needed to organise and deliver the work including its management, monitoring and coordination.

Supervision

Supervisors are a vital part of effective management arrangements and should have the skills, knowledge, training, experience and leadership qualities appropriate to the construction project. Good communication and people management skills on site are important qualities for supervisors.

Information and instruction

▶ Any information or instruction provided should be in simple, clear English (and/or other languages where appropriate) and it should also be set out in logical order (possibly with illustrations).

▶ The use of photographs or diagrams in written communication can also be useful.

▶ The amount of detail should be proportionate to the scale and complexity of the project and the associated health and safety risks.

Site inductions

The principal contractor must ensure that a suitable site induction is provided to every site worker.

Consultation

Proper consultation with the workforce is to be encouraged.

The preparation of the health and safety file

▶ A health and safety file is only required for projects involving more than one contractor.
▶ The client must ensure that the principal designer prepares a health and safety file for their project.
▶ The file must contain information about the current project that is likely to be needed to ensure health and safety during any subsequent work such as maintenance, cleaning, refurbishment or demolition.
▶ The principal contractor must provide the principal designer with any relevant information that needs to be included in the health and safety file.

The duties of domestic clients.

A domestic client is someone who has construction work done on their own home, or the home of a family member, which is not done in connection with a business. A domestic client is not required to carry out the duties placed on commercial clients.

Where the project involves:

a only one contractor, the client duties must instead be carried out by the contractor. In practice, this should involve doing little more than to manage the work to ensure health and safety;

b more than one contractor, the client duties must instead be carried out by the principal contractor as well as the duties they already have as principal contractor. If the domestic client has not appointed a principal contractor then the duties of the client will be carried out by the contractor in control of the construction work.

Much more information on the CDM 2015 Regulations is given in the current edition (5th) of *Introduction to Health and Safety in Construction* by Hughes and Ferrett.

Sources of external construction health and safety information

▶ Construction legislation
▶ Construction publications and reports from the HSE
▶ Reports from trade organisations and construction forums, such as the Construction Industry Training Board (CITB) and Construction Industry Research and Information Association (CIRIA)
▶ The Chartered Institute of Building (CIOB)
▶ The Construction Skills Certification Scheme (CSCS)
▶ The International Labour Organisation (ILO)
▶ CEN, the European Committee for Standardisation
▶ British Standards Institute (BSI)
▶ Construction health and safety forums such as:
 ▷ CIRIA – Construction Industry Research and Information Association
 ▷ CECA – Civil Engineering Contractors Association
 ▷ UKCG – UK Contractors Group.

2.2

Construction site – hazards and risk control

Initial site assessment

Principal health and safety issues on construction sites

These include:

▶ safe place of work
▶ protection against falls
▶ fragile roofs
▶ protection from falling objects
▶ working over water
▶ vehicles and traffic routes
▶ fire and other emergencies
▶ electricity
▶ noise and vibration
▶ health hazards
▶ excavations
▶ demolition.

Initial site assessment

This will involve knowledge of the previous and current use of the site:

- Is it a green field or brown field site?
- History of the site:
 - previously contaminated?
 - old mine workings, shafts, cellars, underground storage and wells?
- A survey of the topography of the site required, including the ground conditions, ground stability and type of ground or soil
- Area and access to the site, particularly for vehicles
- If demolition and/or refurbishment involved, a structural survey should be made of the building (see next section)
- Other activities on the site?
- Will hot work permits be required?
- Presence of overhead and underground services
- Emergency evacuation possible?

Nature of the surroundings to the site:

- the proximity of roads, footpaths, railways, rivers and other waterways
- the details of any fencing or hedging and general vegetation
- the position and description of any trees which need to be felled and those which must be untouched
- the proximity of residential, commercial or industrial properties
- the proximity of schools, colleges, hospitals and playing fields
- the presence of overhead power lines and details of the responsible authority
- the availability of services, such as electricity, telephone lines and water and drainage systems
- any evidence of fly tipping or other unauthorised activity
- measures required to prevent unauthorised access by vehicles.

99

Appropriate general site controls and facilities

1. Site planning

The essential elements are:

- ▶ site layout;
- ▶ location of access and egress points to a site;
- ▶ safe access for pedestrians and vehicles, including emergency services;
- ▶ separation of vehicles and pedestrians on site;
- ▶ suitable areas for parking, loading and unloading;
- ▶ method statements and risk assessments;
- ▶ adequate lighting throughout the site;
- ▶ suitable health and safety, warning or directional signs;
- ▶ areas for site offices, materials and machinery storage;
- ▶ welfare facilities for the workforce;
- ▶ arrangements for site management and control, including inspection procedures and workforce supervision;
- ▶ arrangements for fire and other emergencies and accident reporting;
- ▶ the development of hot work permits, provision of fire-fighting equipment and suitable storage facilities for flammable substances;
- ▶ preparation of site waste management plans; and
- ▶ arrangements for the disposal of any waste, to include:
 - ▷ clear identification
 - ▷ secure storage
 - ▷ transferred only to an authorised person
 - ▷ adequate records kept.

Every construction site must have a first-aid box and eye wash facilities. A small first-aid kit must be given to lone workers on remote sites.

Demolition or refurbishment projects

For these projects, hazards associated with structural collapse, fire, working at height, lifting and carrying and electricity must be

considered together with health hazards from lead, asbestos, vermin and insects.

Under CDM 2015, the client must provide the principal contractor with relevant information about the building structure, including its stability and structural form and any significant design assumptions.

A structural survey before work commences should reduce the risk of unintentional structural collapse. Such a survey should be undertaken by a competent person before any potentially load-bearing parts of a structure are altered. The structural survey should consider:

▶ the age of the structure;
▶ previous use;
▶ type of construction; and
▶ any nearby buildings or structures.

A competent person should decide the method and design of temporary supports. Temporary support provided must be designed, installed and maintained to withstand foreseeable loads and structures should never be overloaded.

See Element 2.12 for more details on demolition.

2. Site preparation

This could involve:

▶ specialist activities – lifting piling or steelwork;
▶ removal or de-branching of trees using chainsaws and a wheeled excavator to remove roots;
▶ removal of contaminated topsoil; and
▶ arrangements to deal with high-voltage overhead power lines.

3. Site security and protecting the public

The client's pre-construction information should include:

▶ project boundaries;
▶ adjacent land use;
▶ access; and
▶ measures to exclude unauthorised people.

101

The protection of the public could involve:

▶ the erection and maintenance of a strong perimeter fence and a lockable gate;
▶ the protection of members of the public, including trespass by children, by:
 ▷ the removal of access ladders
 ▷ providing trench supports
 ▷ covering holes where possible
 ▷ immobilising plant
 ▷ safe stacking of materials
 ▷ locking away dangerous chemicals
 ▷ isolating electrical supplies, and
 ▷ locking off fuel storage tanks;
▶ the provision of warning signs at the perimeter fence and suitable viewing points for the public to observe the construction work safely;
▶ the secure storage of cement, which is a hazardous substance, in a dry store;
▶ the storage of gas cylinders, fuel and other flammable substances in a separate flameproof store; and
▶ the control of excessive noise, dust and mud on the public highway.

4. Arrangements with the client and/or occupier of the premises

These could include:

▶ site rules;
▶ the protection of members of the public, visitors and other employees;
▶ first-aid arrangements;
▶ catering facilities; and
▶ full cooperation from the occupier and his/her staff in the observation of health and safety rules.

For work on a public highway:

▶ ensure the safe passage of vehicles and pedestrians by:
 ▷ the erection of warning signs
 ▷ the use of cones with lead-in and exit tapers and an indication of a safety zone between the cones and the work area
 ▷ the possible use of barriers and lighting for pedestrian safety
 ▷ the possible use of traffic lights or stop/go boards to control traffic; and
 ▷ informing certain neighbours, such as schools, hospitals or factories, of the timing of some aspects of the work.

5. Site induction training topics

These could include:

▶ outline of the project;
▶ site health and safety policy;
▶ site welfare facilities;
▶ site health and safety plans;
▶ those site method statements which are relevant to the workers at the induction;
▶ the findings of any relevant risk assessments;
▶ details of any hazardous operations, such as work at height, site vehicular movement and confined space working;
▶ manual handling procedures and precautions;
▶ accident reporting procedures and first-aid arrangements;
▶ emergency procedures, including fire precautions, emergency telephone numbers and assembly points;
▶ the site rules; and
▶ the managerial and consultation arrangements on site, including the identity of the supervisor.

Summary of a safe place of work for a construction site

▶ secure and locked gates with appropriate notices posted;
▶ a secure and undamaged perimeter fence with appropriate notices posted;

- ▶ all ladders either stored securely or boarded across their rungs;
- ▶ all excavations covered;
- ▶ all mobile plant immobilised, where practicable, and services isolated;
- ▶ secure storage of all inflammable and hazardous substances;
- ▶ visits to local schools to explain the dangers present on a construction site; and
- ▶ if unauthorised entry persists, then security patrols and closed circuit television may need to be considered.

Hazards and control measures for works of a temporary nature

Possible hazards are:

- ▶ slips, trips and falls are a frequent source of injury to members of the host organisation and the public;
- ▶ unfenced excavations;
- ▶ inadequate protection of holes, uneven surfaces, poor reinstatement, trailing leads and cables (especially on stairways), spillage of oils and gravel;
- ▶ poor lighting and ventilation in and around the temporary works;
- ▶ noise and vibration;
- ▶ hazardous substances including dust;
- ▶ falling or flying objects;
- ▶ the accumulation of waste materials;
- ▶ falling or moving objects;
- ▶ poor storage of materials and equipment and other obstructions in public areas, including inadequate control of waste materials, are common causes of accidents;
- ▶ fire and/or explosions;
- ▶ lack of project management and supervision;
- ▶ hazards created outside the site perimeter (for example, unloading materials from a delivery lorry outside the perimeter);
- ▶ presence of asbestos; and
- ▶ demolition work.

Principal control measures are:

▶ the appointment of a Temporary Works Coordinator to oversee the project and to sign it off on completion;

▶ good cooperation between the contractors and host employees;

▶ a comprehensive risk assessment for the project;

▶ good management and supervision of the project;

▶ good communication between clients, contractors and residents;

▶ the provision of welfare facilities including first-aid arrangements; and

▶ the possible segregation of work areas or access to them controlled by physical barriers or warning signs.

Additional controls may include:

▶ covering or fixing of any trailing cables that cross pedestrian areas;

▶ provision of lighting at night and in dark areas;

▶ prompt removal of all waste and rubbish;

▶ protection of vulnerable groups such as the elderly, children and people with certain disabilities;

▶ provision of suitable protection from dust, asbestos fibres, noise and vibration; and

▶ posting of warning signs around the site and at all entrances to the main building.

The risk assessment should include the identification of:

▶ all aspects of the works including the health and safety implications to the host employees and the contractors;

▶ the people at risk from the work including, where appropriate, members of the public; and

▶ any risks to the contractors from the business of the host organisation.

Health, welfare and work environment requirements

Welfare and work environment issues are covered by the Workplace (Health, Safety and Welfare) Regulations.

Welfare facilities include:

▶ the provision of sanitary conveniences;
▶ washing facilities (including showers if necessary);
▶ an adequate supply of drinking water;
▶ accommodation for changing and storage of clothing;
▶ facilities for rest and eating meals at break times;
▶ adequate first-aid provision (an accident book); and
▶ protective clothing against adverse weather conditions.

The rest area should be covered and provided with tables and chairs and a means of heating water.

Hot and cold water and a means of hand drying are the minimum that should be provided on site for drying hands.

The pre-construction information should include the arrangements required for welfare provision. The construction phase must not start unless the client is satisfied that there are arrangements for welfare facilities in place. The facilities must be maintained by the Principal Contractor throughout the life of the project.

Sanitary conveniences and washing facilities

These should include:

▶ the requisite number of water closets, wash stations and urinals for varying sizes of workforce;
▶ special provision for disabled workers;
▶ separate facilities for men and women;
▶ adequate protection from the weather;
▶ a good supply of warm and cold water, soap and towels as close to the sanitary facilities as possible;
▶ in the case of temporary or remote work sites, sufficient chemical closets and washing water in containers;

▶ good ventilation, lighting and regular cleaning; and
▶ only as a last resort should public conveniences be used.

Accommodation for clothing and facilities for changing clothing

Must be:

▶ clean, warm and dry;
▶ well ventilated;
▶ secure; and
▶ it must be ensured that workers' own clothing is not contaminated by any hazardous substances.

Drinking water

▶ Accessible to all the workforce.
▶ The supply of drinking water must be adequate and wholesome.
▶ Normally mains water is used and should be marked as 'drinking water' if water not fit for drinking is also available.
▶ On some sites, particularly during the early stages of construction, it may be necessary to provide bottled water or water in tanks. In this event, the water should be changed regularly and protected from contamination.
▶ Cups or mugs should be available near the tap unless a drinking fountain is provided.

Facilities for rest and eating meals

Must be provided so that:

▶ workers may sit down during break times in areas where they do not need to wear personal protective equipment;
▶ workers can make hot drinks and prepare food;
▶ facilities are provided for pregnant women and nursing mothers to rest; and
▶ food is not contaminated by hazardous substances.

107

The welfare facilities

Should be located:

▶ adjacent to each other; and
▶ as close to the main working area as possible.

Workplace environment

▶ ventilation
▶ heating and temperature (particularly extremes)
▶ lighting:
 ▷ natural light
 ▷ stroboscopic effects
 ▷ special local lighting
 ▷ structural aspects (shadows)
 ▷ atmospheric dust
 ▷ heating effect of lighting
 ▷ lamp and window cleaning
 ▷ emergency lighting
▶ workstations and seating
▶ floors, stairways and traffic routes
▶ translucent or transparent doors should be constructed with safety glass and properly marked to warn pedestrians of their presence
▶ adequate arrangements in place to ensure the safe cleaning of windows and skylights
▶ adequate provisions for the needs of disabled workers.

Exposure to inclement weather

▶ warm clothing in cold weather and
▶ waterproof clothing during wet weather.

Effects of extreme temperature

	Health effects	**Preventative measures**
High temperatures	Heart strain Heat stroke	Supply of drinking water Supply of ventilated air Regular health surveillance Information and training on the hazards
Low temperatures	Frostbite Loss of limbs	Supply of thick, warm (thermal) clothing Provision of hot drinks and external heating Regular health surveillance Information and training on the hazards

Violence at work

Defined as any incident in which a person is abused, threatened or assaulted in circumstances related to their work. High-risk occupations include health and social services, police and fire-fighters, lone and night workers and benefit services.

Workers who are most at risk from violence are those that:

▶ handle money;
▶ provide a service to the public (such as shop workers, teachers and nurses);
▶ are lone workers;
▶ represent authority (such as police, traffic wardens and even school crossing patrols).

Many people resort to violence due to frustration. Common causes of such frustration are the following:

▶ dissatisfaction with a product or service, including the cost;
▶ a perception of being unreasonably penalised over an incident such as car parking; and
▶ a general lack of information following a problem, such as aircraft delays or long delays at hospitals.

109

In over a third of the incidents, the victim believed the offender was under the influence of alcohol, and in nearly a fifth of incidents under the influence of drugs.

Action plan

1 find out if there is a problem
2 decide on what action to take:
 i. quality of service provided
 ii. design of the operating environment
 iii. type of equipment used
 iv. designing the job
3 take the appropriate action
4 check that the action is effective.

Types of security equipment used

▶ access control (swipe cards and simple coded security locks);
▶ closed circuit television (CCTV);
▶ alarms – there are three main types:
 ▷ intruder alarms
 ▷ panic alarms; and
 ▷ personal alarms
▶ radios and pagers (particularly for lone workers); and
▶ mobile phones.

Substance misuse at work

Risks to health and safety of drugs and alcohol

▶ reduce productivity;
▶ increase absenteeism;
▶ increase accidents at work; and, in some cases,
▶ endanger the public.

Alcohol and solvent abuse is a considerable problem when vehicle driving is part of the job, especially if driving is required on public roads.

Controls

For alcohol abuse:

▶ during working hours, there should be no drinking
▶ drinking during break periods should be discouraged
▶ induction training should stress this policy
▶ managers should enforce the policy
▶ posters can also help to communicate the message
▶ the availability of a confidential counselling service.

Drug abuse presents similar problems to those found with alcohol abuse:

▶ absenteeism
▶ reduced productivity
▶ an increase in the risk of accidents
▶ impairments in cognition, perception and motor skills.

Signs of a substance misuse problem

▶ sudden mood changes
▶ unusual irritability or aggression
▶ a tendency to become confused
▶ abnormal fluctuations in concentration and energy
▶ impaired job performance
▶ poor time-keeping
▶ increased short-term sickness leave
▶ a deterioration in relationships with colleagues, customers or management
▶ dishonesty and theft (arising from the need to maintain an expensive habit).

A policy on drug abuse can be established by:

1 **Investigation of the size of the problem.**
2 **Planning actions** – develop an awareness programme for all staff and a special training programme for managers and supervisors.
3 **Taking action** by producing a written policy to include:
 ▶ details of the safeguards to employees
 ▶ the confidentiality given to anyone with a drug problem

▶ the circumstances in which disciplinary and/or reported action will be taken.

4 **Monitoring the policy** by checking for positive changes in the measures made during the initial investigation (improvements in the rates of sickness and accidents).

Safe movement of people on construction sites

Hazards to pedestrians include:

▶ slips, trips and falls on the same level
▶ falls from height
▶ collisions with moving vehicles
▶ being struck by moving, falling or flying objects
▶ striking against fixed or stationary objects
▶ treading on sharp items.

Trip hazards are caused by	Slip hazards are caused by
▶ loose floorboards or carpets	▶ wet or dusty floors
▶ obstructions, low walls, low fixtures on the floor	▶ the spillage of wet or dry substances – oil, water, plastic pellets
▶ cables or trailing leads to portable electrical hand tools and other electrical appliances across walkways	▶ loose mats on slippery floors
▶ raised telephone and electrical sockets	▶ wet and/or icy weather conditions
▶ rugs and mats – particularly when worn or placed on a polished surface	▶ unsuitable footwear, floor coverings or sloping floors
▶ poor housekeeping – obstacles left on walkways, rubbish not removed regularly	
▶ poor lighting levels – particularly near steps or other changes in level	
▶ sloping or uneven floors – particularly where there is poor lighting or no handrails	
▶ unsuitable footwear – shoes with a slippery sole or lack of ankle support	

112

Control strategies for pedestrians

▶ risk assessment to identify suitable controls
▶ slip-resistant surfaces and reflective edges to stairs and kerbs
▶ spillage control and drainage
▶ designated walkways
▶ fencing and guarding, particularly on stairways
▶ use of warning signs
▶ sound storage racking that is inspected and maintained regularly
▶ maintenance of a safe workplace
▶ cleaning and good housekeeping procedures. Signs that warn people that floors are wet are only partially effective. HSE make the following suggestions:
 ▷ physically exclude people from wet cleaning areas by the use of barriers or locking off an area
 ▷ clean during quiet hours
 ▷ clean in sections so there is always a dry path through the area
▶ access and egress
▶ environmental considerations (heating, lighting, noise and dust)
▶ special provision for disabled people
▶ high visibility clothing, appropriate footwear, personal protective equipment (safety harnesses)
▶ training workers in the maintenance of trip-free working areas
▶ regular inspections of work areas by supervisors
▶ information and instruction.

Floor cleaners must be provided with appropriate slip-resistant footwear which can help reduce the risk of slipping.

Permanent staircases are a source of falls from a height and the following design and controls should be considered:

▶ adequate width of the stairway and depth of the tread;
▶ provision of landings and banisters or handrails and intermediate rails;
▶ provision of non-slip surfaces and reflective edging;
▶ adequate lighting;
▶ adequate maintenance;
▶ special or alternative provision for disabled people.

Maintenance activities

The majority of accidents to workers engaged in maintenance work occurred during the following activities:

▶ roof work;
▶ machinery and plant maintenance;
▶ building maintenance including painting work;
▶ maintenance of services (gas, electricity, telecommunications and water);
▶ vehicle repair work; and
▶ window and building cleaning and repair.

The main hazards associated with maintenance work are due to:

▶ electricity – electric shock;
▶ mechanical including entanglement, ejection and unexpected start-up;
▶ hazardous substances – gases, dusts and fume;
▶ physical – noise and vibration;
▶ obstructions of various types;
▶ access – confined spaces and work at height; and
▶ unexpected releases of pressure or other forms of energy.

Hazards and the associated controls to the general public in construction activities

Main hazards:

▶ materials falling from working platforms
▶ striking against fixed or stationary objects
▶ trespass by children
▶ collision with vehicles
▶ operation of cranes and other lifting equipment.

Common control measures:

▶ posting of various warning signs
▶ provision of clearly defined and protected thoroughfares
▶ separate entrances for people and vehicles
▶ provision of 2 m high perimeter fence, with no clearance at ground level

- adequate lighting
- good housekeeping by ensuring that all waste debris and construction materials, particularly timber, is safely stored and/or removed from the site
- the use of appropriate personal protective equipment
- remove all protruding nails or screws or flatten against the timber
- security guards and/or cameras required on sites which are particularly vulnerable to trespassers.

Groups of the public particularly at risk from construction work on public highways are:

- people with disabilities
- the elderly
- children.

Measures to protect site personnel and members of the public while working on public highways:

- temporary traffic controls;
- signs for pedestrians and traffic;
- cones or other barriers to designate the safety zone;
- barriers to protect and guide the public around the site;
- temporary walking surfaces for the public;
- temporary lighting;
- secure and safe storage of working materials;
- arrangements for the movement of vehicles to and from the site;
- provision of high visibility clothing for workers on the site; and
- arrangements for other hazards associated with the work at the site, such as noise, dust and buried services.

Additional control measures for work on public highways:

- means of controlling traffic (traffic lights or stop/go boards);
- the provision of barriers – traffic barrier and/or pedestrian barrier;
- road works signs must be used:
 - ▷ Road Works Ahead
 - ▷ Road Narrows Ahead

115

- ▷ Keep Right and Keep Left signs
- ▷ cones and lamps
- ▷ Information board
- ▷ End sign; and
- ▶ signs placed on footpaths must not create a hazard for pedestrians.

2.3

Vehicle and plant movement – hazards and risk control

Learning outcomes

Explain the hazards and control measures for the safe movement of vehicles/plant within a construction environment, including when using public highways as a workplace ☐

Outline the factors associated with driving at work that increase the risk of an incident and the control measures to reduce work-related driving risks ☐

🔑 **Key revision points**

The hazards and control strategy for safe vehicular operations	☐
The management of vehicle movements on the construction site	☐
The management of occupational driving risks in terms of the driver, vehicle and journey planning	☐
The particular issues in the use of lorries	☐

Safe movement of vehicles and plant within a construction site

Hazards in vehicle operations include:

▶ collisions between pedestrians and vehicles
▶ vehicles crushing feet of pedestrians
▶ people falling from vehicles
▶ people being struck by objects falling from or attached to vehicles
▶ people being struck by or ejected from an overturning vehicle
▶ communication problems between vehicle drivers and employees or members of the public.

Hazards in vehicle operations may be caused by:

▶ poor working practices, such as the lack of regular vehicular safety checks
▶ defective maintenance, steering, brakes, tyres and hydraulic hoses
▶ poor road surfaces and/or poorly drained road surfaces
▶ overloading of vehicles
▶ use of unsuitable vehicles to transport people
▶ inadequate supervision
▶ poor training or lack of refresher training.

Other more general hazards involving pedestrians and vehicles include:

▶ reversing of vehicles, especially inside buildings; pedestrians are most likely to be injured on site by reversing vehicles;

▶ roadways too narrow with insufficient safe parking areas;

▶ roadways poorly marked out and inappropriate or unfamiliar signs used;

▶ too few pedestrian crossing points;

▶ the non-separation of pedestrians and vehicles;

▶ lack of barriers along roadways;

▶ lack of directional and other signs;

▶ poor visibility due to load impeding view, blind corners or inadequate vehicular lighting or mirrors;

▶ poor environmental factors, such as lighting, dust and noise;

▶ ill-defined speed limits and/or speed limits which are not enforced; and

▶ vehicles used by untrained and/or unauthorised personnel.

Hazards from mobile work equipment

Accidents involving mobile work equipment often arise from one or more of the following events:

▶ poor maintenance with defective brakes, tyres and steering;

▶ poor visibility;

▶ operating on rough ground or steep gradients;

▶ carrying of passengers without the proper accommodation for them;

▶ people being ejected as the vehicle overturns and/or being crushed by it;

▶ being struck by a vehicle or an attachment;

▶ lack of driver training or experience;

▶ poor management procedures, training and supervision;

▶ collision with other vehicles;

▶ overloading of vehicles;

▶ general vehicle movements and parking;

▶ dangerous occurrences or other emergency incidents; and

▶ access and egress from the buildings.

119

Control strategies for safe vehicle operations

- ▶ risk assessment
- ▶ designated marked traffic routes with good visibility and well-designed loading and storage areas
- ▶ enforced speed limits and/or the fitting of speed governors
- ▶ audible warning of approach signals, particularly for reversing
- ▶ separation of pedestrians and vehicles
- ▶ use of one-way systems
- ▶ the issue of suitable high visibility clothing and other personal protective equipment to drivers and pedestrians in vehicle operating areas
- ▶ separate site access gates for pedestrians and vehicles
- ▶ use of pedestrian crossings on vehicular routes
- ▶ mirrors on blind corners
- ▶ identification of recognised parking and non-parking areas
- ▶ include vehicle safety in employee induction training
- ▶ driver training and refresher training given by competent persons
- ▶ the driver is responsible for the security of the load on a vehicle
- ▶ the provision of driver protection by the use of:
 - ▷ seat belts
 - ▷ FOPS (falling-object protective structures)
 - ▷ ROPS or TOPS (roll- or tip-over protective structures)
 - ▷ banksmen during reversing operations
- ▶ effective vehicular maintenance procedures
- ▶ use of daily vehicular inspection checklists to include emergency systems, engine oil level and hydraulic fluid level
- ▶ environmental considerations (road surfaces, gradients and changes in road level, lighting, visibility).

The management of vehicle movements involves:

- ▶ a designated management system and a code of practice for all drivers, who must be competent
- ▶ an effective line management organisation to ensure supervision of all vehicular activities
- ▶ a documented preventative maintenance programme with all work recorded

120

▶ regular inspections and, in some cases, thorough examinations by competent persons

▶ ensuring that all vehicles are fitted with suitable driver protection – seat belts, ROPS

▶ ensuring that all vehicles are fitted with reversing warning systems

▶ suitable fire precautions, particularly in battery charging areas.

Hazards and controls of vehicles on construction sites

Common construction vehicles:

▶ site dumper trucks
▶ fork-lift trucks with telescopic handlers
▶ all-terrain and rough terrain vehicles
▶ excavators
▶ tipper lorries
▶ various types of earth moving equipment.

Main hazards:

▶ overturning;
▶ collision with structures on the site, pedestrians or other site vehicles;
▶ non-segregation of pedestrians and vehicles;
▶ uneven road surfaces;
▶ inclines which are too steep;
▶ falling materials;
▶ noise;
▶ dust;
▶ poor maintenance.

The methods of construction can also present hazards (excavations, scaffolds and falsework).

Key controls are:

▶ provision of safe vehicular routes – clear routes should be set out across the site. These routes should avoid sharp bends, blind corners (unless suitably placed mirrors are located at the corner),

121

narrow gaps, low head room and adverse cambers or steep gradients;
- ▶ road surfaces inspected and maintained regularly;
- ▶ the use of one-way systems;
- ▶ introduction and enforcement of speed limits;
- ▶ adequate lighting at road junctions;
- ▶ separate entrances and exits to the site with adequate turning room;
- ▶ the use of clear signs;
- ▶ prevention of falls into excavations by the erection of strong barriers;
- ▶ separation of pedestrians and vehicles;
- ▶ the safety of members of the public must be considered, particularly where vehicles cross public footpaths;
- ▶ restriction of the carriage of people on site;
- ▶ loading and storage areas need to be well designed;
- ▶ only properly trained drivers allowed to drive vehicles;
- ▶ banksmen should be used to direct lorries and excavators;
- ▶ rollover and falling-object protection;
- ▶ driver restraint systems;
- ▶ good maintenance of vehicles (and daily checks);
- ▶ vehicles fitted with flashing beacons or reversing alarms;
- ▶ vehicles must not be overloaded; and
- ▶ vehicles must be sheeted when loaded.

Driving at work

A risk assessment should be undertaken to cover all aspects of driving at work. This should include the use of alternative means of transport other than driving.

Factors affecting occupational road risks

The driver

- ▶ competency issues including:
 - ▷ validity of the driving licence
 - ▷ awareness of the company policy on work-related road safety

▷ general driving experience
▶ health and fitness, in particular eyesight
▶ level of training received.

The vehicle

▶ its suitability;
▶ its condition, particularly seat belts and head restraints;
▶ its past maintenance schedule;
▶ the ergonomics of the driving position;
▶ any load should be stacked safely so that it cannot move during the journey. There must also be satisfactory arrangements for handling the load at either end of the journey;
▶ the vehicle breakdown procedure.

The journey

▶ route to be undertaken, including the amount of traffic, the time of day when the journey is made and the sufficiency of the time allowed for its completion;
▶ the type of roads being used, whether motorway, minor roads or those involving city centre congestion, and their condition;
▶ schedules need to take sufficient account of busy periods on roads;
▶ driving excessive distances without appropriate breaks; and
▶ adverse weather conditions, such as snow, ice, heavy rain and high winds.

Managing occupational road risks

An effective policy to manage occupational road risk should address important and relevant aspects of the Highway Code, the possible enforcement action by the authorities and the following issues:

For the driver

▶ all drivers must have a current and valid driving licence;
▶ details of a prospective driver's previous driving experience should be examined and recorded;

123

- ▶ driver training and health;
- ▶ regular eyesight tests;
- ▶ staff should not drive, or undertake other duties, while taking a course of medicine that might impair their judgement;
- ▶ some prescribed and over-the-counter drugs and medicines can affect driver awareness and speed of reaction. Drivers using such medicines must check with a doctor or pharmacist to ensure that it is safe to drive;
- ▶ ensure that drivers have necessary communication equipment (mobile phones, GPS, personal alarms);
- ▶ encourage drivers.to regularly report back to colleagues in the base office;
- ▶ the monitoring of driver performance;
- ▶ the number of recorded incidents;
- ▶ all speed limits must be observed;
- ▶ the determination and recording of driving hours;
- ▶ routine driver safety checks such as those on lights, tyres and wheel fixings;
- ▶ drivers and passengers are adequately protected in the event of an incident. Crash helmets and protective clothing for those who ride motorcycles and other two-wheeled vehicles should be of the appropriate colour and standard;
- ▶ drivers of Heavy Goods Vehicles (HGVs) must have the appropriate driving licence and medical certificate; and
- ▶ disciplinary action taken in the event of non-compliance with the agreed in-house rules and standards.

For the vehicle

- ▶ all vehicles must carry comprehensive insurance for use at work;
- ▶ all vehicles should be fit for purpose;
- ▶ new or replacement vehicles should be suitable for both driving and the health and safety of the public;
- ▶ all vehicles should be regularly serviced and maintained – particularly important when pool cars are used, because pool car users often assume another user is checking on maintenance and the MOT;

▶ the safety critical systems (brakes, steering and tyres) need to be properly maintained;

▶ if the vehicle is leased and serviced by the leasing company, a system should be in place to confirm that servicing is being done to a reasonable standard;

▶ each vehicle should have a handbook available for the driver;

▶ there should be a defect reporting system for drivers to use if they consider their vehicle is unsafe; and

▶ if privately owned vehicles are used for work, they should be insured for business use and, if necessary, have an appropriate MOT certificate test.

For lorries

▶ the policy should address issues associated with the loads;

▶ if the load is hazardous, emergency procedures (and possibly equipment) must be in place and the driver trained in those procedures;

▶ the load should be stacked safely in the lorry so that it cannot move during the journey;

▶ there must also be satisfactory arrangements for handling the load at either end of the journey;

▶ the driver must be aware of the height of the vehicle, both laden and empty; and

▶ where tachographs are carried, they should be checked regularly to make sure that drivers are not driving for long periods without a break.

For journey planning

▶ journey times must include statutory breaks and allow for possible delays;

▶ motorways are the safest roads – minor roads are suitable for cars, but are less safe and could present difficulties for larger vehicles;

▶ overhead restrictions, for example bridges, tunnels and other hazards such as level crossings, may present dangers for long and/or high vehicles;

- ▶ sufficient consideration will need to be given to adverse weather conditions; and
- ▶ no alcohol or recreational drugs must be consumed during the day of the journey until the journey is completed. Only minimal amounts of alcohol should be consumed on the day before a journey is to be made.

Musculoskeletal hazards and risk control

Learning outcomes	
Explain work processes and practices that may contribute to musculoskeletal disorders, work-related upper limb disorders and the appropriate control measures	☐
Explain the hazards and control measures which should be considered when assessing risks from manual handling activities	☐
Explain the hazards and control measures to reduce the risk in the use of lifting and moving equipment, with specific reference to manual and mechanically operated load moving equipment	☐

🔑 Key revision points

Musculoskeletal health issues and ill-health prevention strategies	☐
The types of injuries resulting from manual handling operations	☐
The development of manual handling risk assessments and effective control measures	☐
Safety in the use of lifting equipment and an understanding of the different types of mechanical handling and lifting equipment	☐
Hazards and controls associated with fork-lift trucks, hoists and cranes	☐
The legal requirements for the inspection and examination of lifting equipment	☐

Musculoskeletal disorders and work-related upper limb disorders

Ergonomics – interaction between workers, their work and their environment. It involves knowledge of their physical and mental capabilities in addition to their understanding of the job.

Musculoskeletal disorders (MSDs) affect muscles, joints, tendons, ligaments and nerves. MSDs usually affect the back, neck, shoulders and upper limbs (termed work-related upper limb disorders – WRULDs).

Ill-health effects due to poor ergonomics

Work-related upper limb disorders (WRULDs):

▶ Tenosynovitis
▶ Repetitive strain injury (RSI)
▶ Carpal tunnel syndrome
▶ Frozen shoulder.

128

Often caused by repetitive operations – keyboard operations, brick laying, kerb laying, moving and fixing of plasterboard, erecting and dismantling of scaffolds and digging.

Display Screen Equipment (DSE)

Requirements of the Health and Safety (Display Screen Equipment) Regulations:

▶ Suitable and sufficient risk assessment
▶ Workstation compliance with minimum specifications given in the Regulations
▶ Adequate breaks in work programme
▶ Free sight test if required
▶ Information and training.

Ill-health hazards from DSE work:

▶ Musculoskeletal problems
▶ Visual problems
▶ Psychological problems.

DSE risk assessment

The DSE risk assessment should consider the following factors:

▶ the height and adjustability of the monitor
▶ the adjustability of the keyboard, the suitability of the mouse and the provision of wrist support
▶ the stability and adjustability of the DSE user's chair
▶ the provision of ample foot room and suitable foot support
▶ the effect of any lighting and window glare at the workstation
▶ the storage of materials around the workstation
▶ the safety of trailing cables, plugs and sockets
▶ environmental issues – noise, heating, humidity and draughts.

Manual handling hazards, risks and control measures

Manual handling can involve any load movement by human effort only (lifting, pushing, pulling, carrying or supporting).

129

Requirements of the Manual Handling Operations Regulations

- ▶ to avoid manual handling, if possible;
- ▶ to mechanise or automate the lifting process, if possible;
- ▶ if unavoidable, undertake a risk assessment;
- ▶ instigate risk controls and review; and
- ▶ give employees information on loads.

Manual handling hazards are:

- ▶ lifting a load which is too heavy and/or cumbersome
- ▶ poor posture and/or poor lifting technique
- ▶ dropping the load on the foot
- ▶ lifting sharp-edged or hot loads.

Injuries caused by manual handling include:

- ▶ muscular sprains or strains
- ▶ back injuries
- ▶ trapped nerve
- ▶ hernias
- ▶ cuts, bruises and abrasions
- ▶ fractures
- ▶ work-related upper limb disorders (WRULDs)
- ▶ rheumatism.

Manual handling risk assessment requires the use of mechanical aids; if this is not possible, then the assessment of:

- ▶ the task
- ▶ the individual
- ▶ the load
- ▶ the working environment.

Manual handling training includes:

- ▶ the types of injury;
- ▶ the manual handling assessment findings;
- ▶ potentially hazardous manual handling operations;
- ▶ the correct use of manually operated load moving equipment, such as sack and pallet trucks;
- ▶ the correct use of mechanical aids;

▶ the correct use of personal protective equipment;

▶ good housekeeping issues;

▶ the factors that can affect an individual; and

▶ a good lifting technique, such as:

1 Check that suitable clothing (including gloves, if required) and footwear are being worn. Make an approximate assessment of the weight of the load and decide whether to lift it alone. Check to see whether one side of the load is heavier than the other and ensure that the heaviest side of the load is closest to the body.

2 Place the feet apart and adopt a good posture by bending the knees and keeping the back straight.

3 Get a firm grip and hold the load as close as possible to the body.

4 At the start of the lift, moderate flexion (slight bending) of the back, hips and knees is preferable to fully flexing the back (stooping) or the hips and knees (squatting).

5 Do not flex the spine any further during the lift.

6 After ensuring that full visibility is available, move forward without twisting the trunk.

7 Set the load down either at waist level or by lowering it first to knee level and then to the floor.

8 Manoeuvre the load to its final position after it has been set down.

Handling kerb stones:

The hierarchy of control measures is:

1 eliminate manual lifting at the design stage, perhaps by dispensing with the need for a kerb

2 total mechanisation, by ensuring that kerb stones are always lifted and laid mechanically

3 partial mechanisation, by ensuring that the maximum amount of kerb handling is done mechanically

4 manual handling where it is not possible to use any of the above methods.

Alternatives to the traditional laying of kerb stones – plastic kerb stones, hollow kerb stones or slip-form kerb stones. The laying

of heavy (over 20 kg) blocks can also lead to musculoskeletal problems.

Lifting and moving equipment

Types of manually operated load handling equipment

▶ simple tools
▶ wheelbarrows
▶ trucks and trolleys
▶ roller tracks and chutes
▶ pallet trucks
▶ conveyors
▶ various types of hoists that can be used to lift people as well as other loads.

Hazards associated with manually operated load handling equipment

▶ incorrect use, such as overloading or attempting to carry unstable loads
▶ manual handling hazards due to pushing or pulling a truck or trolley
▶ lack of maintenance
▶ injuries to pedestrians using the same walkway.

Precautions with the use of manually operated load handling equipment

▶ when handling aids are being selected, the subsequent user should be consulted whenever possible;
▶ moving and handling tasks are often made easier by good design;
▶ a safe system of work should be in place;
▶ all operators of mechanical handling equipment must be properly trained in its use and supervised while they are using it;
▶ sufficient room available to manoeuvre the equipment easily;
▶ adequate visibility and lighting available;
▶ the floor is in a stable condition;
▶ regular safety checks to identify any faults with the equipment;

- a regular maintenance schedule for the equipment;
- the proposed use will be within the safe working load of the equipment;
- the lifting equipment is CE-marked; and
- advice is sought from the suppliers/hirers on its suitability for the proposed task and any maintenance requirements.

Safety in the use of mechanically operated lifting equipment

Positioning and installation of lifting equipment

Risks during lifting operations should be reduced by:

- avoiding the equipment or its load striking a person;
- avoiding a load drifting, falling freely or being released unintentionally;
- avoiding the need to lift loads over people;
- stopping safely in the event of a power failure;
- where possible, enclosing the path of the load with suitable and substantial interlocked gates.

The organisation of lifting operations

Every lifting operation, which is lifting or lowering a load, shall be:

- properly planned by a competent person;
- appropriately supervised; and
- performed in a safe manner.

Summary of the requirements for lifting operations

There are four general requirements for all lifting operations:

- strong, stable and suitable lifting equipment;
- the equipment should be positioned and installed correctly;
- the equipment should be visibly marked with the safe working load; and
- lifting operations must be planned, supervised and performed in a safe manner by competent people.

133

Lift plans

A lift plan identifies the ways in which the risks involved in a lifting operation can be eliminated or controlled and involves:

1 initial planning to ensure that the lifting equipment provided is suitable for the task; and
2 the planning of individual lifting operations so that they can be performed safely with the lifting equipment provided.

The details of the plan depend on the type of lifting equipment.

Types of mechanical handling and lifting equipment

Conveyor

There are three types of conveyor – belt, roller and screw.

Typical hazards include:

- ▶ in-running nips
- ▶ entanglement
- ▶ loads falling from conveyor
- ▶ impact against overhead systems
- ▶ contact hazards – sharp edges
- ▶ manual handling hazards
- ▶ noise and vibration hazards.

Elevator

An elevator transports goods between floors. The most common hazard is caused by loads falling from the elevator. Also there are manual handling hazards at either end of the elevator.

Fork-lift trucks

Hazards include:

- ▶ overturning
- ▶ overloading
- ▶ collisions with pedestrians or other vehicles or structures

▶ the silent operation of electrically powered fork-lift trucks makes pedestrians unaware of their presence
▶ uneven road surface
▶ overhead obstructions
▶ loss of load
▶ failure of the hydraulic system
▶ inadequate maintenance
▶ lack of driver training
▶ use as a work platform and/or carrying passengers
▶ speeding
▶ poor vision around load
▶ dangerous stacking technique, particularly on warehouse racking
▶ fire – when battery charging or refuelling.

There are also the following physical hazards:

▶ noise
▶ exhaust fumes
▶ vibrations
▶ manual handling
▶ ergonomic – musculoskeletal injuries due to soft tyres, expansion joints.

Driver daily checks:

▶ tyres and tyre pressures
▶ brakes
▶ reversing horn and light
▶ mirrors
▶ secure and properly adjusted seat
▶ correct fluid levels
▶ correct working of all lifting and tilting systems.

Lifting equipment on construction sites

Types of equipment:

▶ excavators
▶ telehandlers
▶ rough-terrain vehicles

- dumper trucks
- hoists.

Causes of accidents:

- overturning on slopes and at the edges of excavations;
- poorly maintained braking systems;
- driver error due to lack of training and/or inexperience.

The following site controls for lifting equipment should be in place:

- designated traffic routes and signs;
- speed limits;
- stop blocks used when the vehicle is stationary;
- proper inspection and maintenance procedures;
- procedures for starting, loading and unloading the vehicle;
- provision of rollover protective structures (ROPS) and seat restraints;
- provision of falling-object protective structures (FOPS);
- visual and audible warning of approach;
- if necessary, hearing protection;
- before a lifting operation starts, a method statement of the sequence of operations must be documented; and
- before lifting equipment is used, check that it is free from defects and has a current inspection certificate.

Cranes

Principles of safe operation

- brief inspection prior to each use (including lifting tackle);
- loads not left suspended when crane not in use;
- prior to lift, ensure that nobody can be struck by load or crane;
- the load must not exceed the safe working load (SWL);
- never carry loads over people;
- good visibility and communications;
- any banksmen or signallers used during the lifting operation must be trained and competent;
- only lift loads vertically – do not drag load;
- travel with load as close to the ground as possible; and
- switch off power to crane when unattended.

136

Mobile jib cranes:

▶ plan each lift (maximum load and radius of operation);
▶ identify overhead obstructions and hazards;
▶ assess load-bearing capacity of ground; and
▶ if fitted, outriggers should be used.

Tower cranes

Typical causes of serious incidents with tower cranes include:

▶ mechanical failure of the brake or lifting ram;
▶ overturn of the crane;
▶ jib collapse;
▶ a load or dropped load striking a worker;
▶ sling failure.

Reasons for crane failure:

▶ overloading
▶ poor slinging of load
▶ overturning
▶ collision with another structure or overhead power lines
▶ foundation failure
▶ structural failure of the crane
▶ operator error
▶ lack of maintenance and/or regular inspections.

During lifting operations, ensure that:

▶ driver has good visibility;
▶ there are no pedestrians below load and barriers are in place; and
▶ audible warning given prior to lifting operation.

Lifts, hoists and items of lifting tackle

A **lift** or **hoist** incorporates a platform or cage. Its movement is restricted by guides and it may carry passengers and/or goods alone.

They are required to:

▶ be of sound mechanical construction;

137

▶ have interlocking doors or gates that must be completely closed before the lift or hoist moves;

▶ be fitted, if carrying passengers, with an automatic braking system to prevent overrunning and a safety device to support the lift in the event of suspension rope failure;

▶ be rigorously maintained by competent persons; and

▶ protect others during maintenance operations from falling down the lift shaft and other hazards.

Other **items of lifting tackle** include chain slings and hooks, wire and fibre rope slings, eyebolts and shackles.

Important points include:

▶ loads must be properly secured and balanced in slings;

▶ lifting hooks should be checked for wear and hook distortion;

▶ shackles and eyebolts must be correctly tightened;

▶ slings must be checked for any damage before use and only used by competent people;

▶ training and instruction should be given in the use of lifting tackle;

▶ regular inspections of tackle should be made in addition to the mandatory thorough examinations; and

▶ lifting items should be carefully stored between uses.

The equipment should be inspected at suitable intervals between thorough examinations. The frequency and the extent of the inspection are determined by the level of risk presented by the lifting equipment. A report or record should be made of the inspection, which should be kept until the next inspection. Unless stated otherwise, lifts and hoists should be inspected every week.

Passenger lifts

Passenger-carrying lifts must be fitted with an automatic braking system to prevent overrunning, at least two suspension ropes, each capable alone of supporting the maximum working load, and a safety device which could support the lift in the event of suspension rope failure. Maintenance procedures must be rigorous, recorded and only undertaken by competent persons. It is very important that a safe system of work is employed during maintenance operations to

protect others, such as members of the public, from falling down the lift shaft and other hazards.

The following additional issues are relevant to the safe operation of passenger lifts:

▶ An emergency mushroom stop button should be fitted in the lift.

▶ A direct telephone line is required if the internal alarm sounder/siren cannot be heard by rescue staff in the building.

▶ A risk assessment for possible injuries to the lift passengers should be undertaken on the design, sensitivity and performance of the door edges. The lift inspector should have the relevant standard to help with this.

▶ A lift stop switch should be available in the pit as well as the motor room at the top of the lift.

▶ The lift beam should be inspected by the lift inspector – the safe working load should be stated on it to prevent overload.

▶ If there are any moving or revolving mechanisms accessible by lift technicians performing their maintenance tasks, then they should be protected from inadvertent contact by suitable guards.

▶ It is advisable to locate a fire extinguisher inside the access door of the motor room although this is not a statutory requirement.

▶ Emergency door release devices must not be routinely operated during non-emergency situations. The emergency keys are intended to allow emergency access to the lifting platform in the event of people becoming trapped and should be under strict control.

▶ Emergency unlocking should be undertaken only in exceptional circumstances and by suitably trained and authorised people.

▶ Safe working procedures and arrangements should be in place setting out what to do in the event of an emergency or failure. For example, how to deal with trapped people and the arrangements for repairing faults.

▶ All lifts and lifting platforms must be inspected, serviced and maintained. Where the lift is used for work purposes, it must be thoroughly examined by a competent person.

Statutory examination of lifting equipment

The Lifting Operations and Lifting Equipment Regulations (LOLER) require inspections and thorough examinations.

An **inspection** is performed by a competent person appointed by the employer and is used to identify whether the equipment can be operated, adjusted and maintained safely.

The following points apply to inspections:

▶ they take place between thorough examinations
▶ the frequency depends on risk assessment
▶ the report or record is kept until next inspection
▶ lifts and hoists should be inspected every week.

A **thorough examination** is a detailed examination of lifting equipment by a competent person who is independent of the employer. The examination is usually made in accordance with a written scheme and a report (within 28 days) is submitted to the employer. The report must be kept for 2 years or until the next thorough examination (whichever is longer).

A thorough examination should be undertaken at the following times:

▶ before first use;
▶ after it has been assembled at a new location;
▶ at least every 6 months for lifting equipment or lifting accessory used for lifting persons;
▶ at least every 12 months for all other lifting equipment;
▶ in accordance with a particular examination scheme drawn up by an independent competent person; and
▶ each time that exceptional circumstances, which are likely to jeopardise the safety of the lifting equipment, have occurred (such as severe weather).

Work equipment hazards and risk control

Learning outcomes	
Outline the general requirements for work equipment	☐
Outline the hazards and controls for hand-held tools, both powered and non-powered	☐
Describe the main mechanical and non-mechanical hazards of machinery	☐
Explain the main control measures for reducing risks from common construction machinery hazards	☐

🔑 **Key revision points**

The definition of work equipment	☐
The legal requirements for the supply and use of work equipment	☐
The safe use and maintenance of work equipment	☐
The hazards and controls related to hand-held tools	☐
Examples of mechanical and non-mechanical machinery hazards	☐
Hierarchy of measures to reduce the risks of injury from work equipment	☐
Practical safeguards and their application to a range of machines	☐
The construction requirements of guards	☐

General requirements of work equipment

There are **TWO** types of law affecting work equipment:

▶ equipment supply law (designed to prevent barriers to trade) – The Supply of Machinery (Safety) Regulations, and
▶ equipment user law (designed to protect people at work) – The Provision and Use of Work Equipment Regulations (PUWER).

Supply of work equipment

Most new work equipment should have 'CE' marking when purchased. The 'CE' mark requires manufacturers to:

▶ identify the health and safety hazards likely to be present when the machine is used;
▶ assess the associated risks;
▶ remove the significant hazards at the design stage or, if that is not possible,

142

▶ provide safeguards (such as guarding or noise enclosures) or, if that is not possible,
▶ use warning signs on the machine to warn of hazards.

Additional requirements of equipment manufacturers:

▶ retain related information in a technical file;
▶ fix a CE mark to the equipment – indicates compliance with all relevant supply laws;
▶ issue a 'Declaration of Conformity' for the equipment.

The **Declaration of Conformity** states that the equipment complies with all relevant health and safety requirements. The declaration must:

▶ state the name and address of the manufacturer or importer into the EU;
▶ contain a description of the equipment, its make, type and serial number;
▶ indicate all relevant European Directives with which the machinery complies;
▶ state details of any notified body that has been involved;
▶ specify which standards have been used in the manufacture (if any);
▶ be signed by a person with authority to do so; and
▶ provide the purchaser with instructions on the safe installation, use and maintenance of the equipment.

Use of work equipment

Definition of work equipment

Any equipment used by an employee at work, such as:

▶ hand tools – hammers, knives, spanners
▶ machines – photocopiers, drilling machines, vehicles
▶ apparatus – laboratory equipment
▶ lifting equipment – fork-lift trucks, hoists
▶ other equipment – ladders, vacuum cleaners
▶ installations – scaffolds, sound enclosures.

Uses of work equipment include:

▶ starting and stopping
▶ repair and maintenance

- ▶ modification and servicing
- ▶ cleaning
- ▶ transporting.

Management duties

Work equipment must:

- ▶ be suitable for its purpose of use and only used for specified operations, and
- ▶ suitable in terms of initial integrity and place of use;
- ▶ be selected with the health and safety of the user in mind;
- ▶ be maintained in an efficient working order and in good repair;
- ▶ be inspected at regular intervals;
- ▶ be restricted in use to designated persons when it has specific hazards;
- ▶ be used together with appropriate information, instruction and training; and
- ▶ conform to EU requirements.

Other duties include:

- ▶ the guarding of dangerous parts of work equipment;
- ▶ the provision of stop and emergency stop controls;
- ▶ ensuring that the equipment is stable;
- ▶ the lighting around the equipment is suitable and sufficient; and
- ▶ suitable warning markings or devices, such as flashing lights, are fitted.

Hierarchy of risk control for work equipment

1 eliminate the risks
2 control risks by physical methods (such as guarding)
3 software measures (such as safe systems of work).

Information, instruction and training

Should include:

- ▶ the significant health and safety issues with the equipment;
- ▶ any limitations or problems with the use of the equipment;

▶ safe methods to deal with these problems;
▶ details of any residual risks associated with the equipment;
▶ details of safe working procedures; and
▶ the location of the manufacturer's manual or guidance.

Induction training will be required on recruitment but refresher training will also be needed when:

▶ jobs are changed, particularly if the level of risk increases;
▶ new technology or equipment is introduced;
▶ the system of work is modified;
▶ there are legislative changes; or
▶ periodically as a refresher course.

Types of maintenance

1 Preventative planned maintenance
2 Condition-based maintenance
3 Breakdown-based maintenance.

Hazards associated with maintenance

▶ lack of competence and/or training;
▶ equipment not made safe before maintenance;
▶ lack of permit to work; or
▶ use of incorrect tools or unsafe equipment.

Inspection of work equipment

▶ after installation or when put into service for the first time;
▶ after assembly at a new site or in a new location; and thereafter
▶ at suitable intervals; and
▶ each time exceptional circumstances occur that could affect safety.

Equipment controls

The equipment should have:

▶ operating controls easily reached and operated and with adequate markings and warning signs;

- ▶ no obstructions or debris around it;
- ▶ emergency controls (red emergency stop buttons); and
- ▶ stability during use.

Other issues

- ▶ the level and quality of general and local lighting
- ▶ clear and durable markings on equipment.

Statutory requirements for examination of boilers and air receivers

- ▶ supplied with correct written information and markings;
- ▶ properly installed;
- ▶ used within its operating limits;
- ▶ written scheme for periodic examination certified by a competent person;
- ▶ examined by a competent person in accordance with the written scheme; and
- ▶ a report of the periodic examination held on file and any required actions undertaken.

Responsibilities of users

- ▶ take care for the health and safety of themselves and others who may be affected by their acts or omissions;
- ▶ cooperate with their employer, as far as is necessary, to ensure compliance with any statutory health and safety duty;
- ▶ not to interfere with or deliberately misuse anything provided, in accordance with health and safety legislation, to further health and safety at work;
- ▶ use any equipment or substance in accordance with any training or instruction given by the employer;
- ▶ keep the area around the machine clear and free of obstructions;
- ▶ wear suitable clothing, eye protection and footwear;
- ▶ report to the employer any serious or imminent danger; and
- ▶ report any shortcomings in the employer's protective health and safety arrangements.

146

Hand tools

Hand-held tools

Typical hazards include:

▶ broken handles
▶ incorrect use of knives, saws and chisels
▶ tools that slip
▶ splayed spanners
▶ flying particles
▶ electrocution and/or burns.

Controls:

▶ suitable for purpose (e.g. insulated tools for electricians)
▶ inspection on a regular basis
▶ training for all hand tool users.

Hand-held power tools

Examples of hand-held power tools:

▶ pneumatic drills/chisels
▶ electric drills
▶ disc-cutters/cut-off saws
▶ sanders
▶ cartridges
▶ pneumatic nail guns
▶ chainsaws.

Hazards include:

▶ entanglement
▶ flying particles (eye injury)
▶ contact with cutting edges
▶ striking hidden/buried services (electrical cables, gas mains)
▶ manual handling injuries
▶ hand–arm vibration syndrome
▶ noise levels
▶ dust levels
▶ trailing leads
▶ fire and/or explosion.

Typical controls:

- read operating instructions before initial use;
- maintain a clean, tidy and well-lit working area;
- never use power tools near water, combustible fluids or gases;
- store tools in a dry, safe and secure place;
- never overload tools;
- always use the correct tool and/or attachment for heavy jobs;
- always wear suitable work clothes and appropriate personal protective equipment;
- ensure that the work piece is securely held in clamps or a vice;
- dust levels from concrete cutting operations can be controlled by wet cutting and/or by fitting a dust collector to the machine;
- maintain tools with care, keep them clean and sharp and inspect cables regularly;
- always check that the power switch is turned off before connecting the power cable; and
- use only tool accessories and attachments recommended by the tool manufacturer.

The use of chainsaws is very hazardous because:

- there is no guard on the cutting chain
- they are heavy
- they are noisy.

Machinery hazards

Mechanical machinery hazards

- crushing
- shearing
- cutting or severing
- entanglement
- drawing-in or trapping (in-running nips)
- impact
- stabbing or puncture

▶ friction or abrasion
▶ high-pressure fluid ejection.

Non-mechanical machinery hazards

These include:

▶ access to the machinery
▶ slips, trips and falls
▶ falling and moving objects from the machinery
▶ obstructions and projections
▶ manual handling and lifting
▶ electricity (shock, burns or fire)
▶ fire and explosion
▶ noise and vibration
▶ high-pressure ejection from hydraulic hose leaks
▶ high/low temperature
▶ dust, fume or mist
▶ radiation – ionising and non-ionising
▶ lack of adequate lighting.

Examples of equipment with machinery hazards:

▶ Photocopier
▶ Document shredder
▶ Bench-top grinding machine
▶ Pedestal drill
▶ Bench-mounted circular saw
▶ Hand-fed power planer
▶ Spindle moulding machine
▶ Compressor
▶ Cement mixer
▶ Plate compactor
▶ Ground consolidation equipment
▶ Circular saw
▶ Road marking equipment
▶ Electrical generator.

Control measures for reducing risks from mechanical hazards

Practical safeguarding

Defined by the legal definition of 'practical' – if it is technically possible to guard an item of equipment then it should be guarded irrespective of cost and convenience.

Levels of protection or hierarchy of measures

(Regulation 11 of PUWER)

1 Fixed enclosing guarding
2 Other guards or protection devices (interlocked, trip, etc.)
3 Protection appliances (jigs, push sticks, etc.)
4 Provision of information, instruction, training and supervision.

Types of guard

Guard	Example
Fixed	around a fan
Adjustable	around a drill bit on a pedestal drill
Interlocking	on the gates of a passenger lift, microwave oven
Trip device	photoelectric cell in front of a guillotine
Two-handed control	crimping machine
Hold-to-run	Tube train driver

Apply these types of guard to the 14 pieces of equipment listed above.

Factors in guard construction

▶ Strength
▶ Weight and size
▶ Compatibility with process
▶ Hygiene
▶ Visibility
▶ Noise attenuation
▶ Free flow of air (ventilation and/or cooling)
▶ Free of sharp edges
▶ Maintenance and cleaning
▶ Openings.

150

2.6

Electrical safety

> **Key revision points**

The basic electrical principles and definitions ☐

The causes and treatment of electric shock ☐

Other electrical hazards and injuries and their associated controls ☐

The selection, suitability, maintenance and inspection of electrical equipment ☐

The advantages and limitations of protective systems ☐

The hazards and controls associated with overhead electrical services ☐

The application, advantages and limitations of portable electrical appliance testing ☐

Hazards and risks associated with the use of electricity

Electrical principles and definitions

- ▶ voltage (volts V)
- ▶ current (amps A)
- ▶ resistance (ohms Ω)
- ▶ Ohm's law – $V = I \times R$
- ▶ Power $P = V \times I$ watts
- ▶ conductors – electricity will flow easily
- ▶ insulators – very poor conductors
- ▶ earthing – connection to earth
- ▶ short circuit – direct flow to earth
- ▶ low voltage – not exceeding 600 V
- ▶ high voltage – exceeding 600 V
- ▶ mains voltage – 220/240 V.

Electrical hazards and injuries

- ▶ electric shock
- ▶ electric burns
- ▶ electric fires and explosions
- ▶ arcing
- ▶ portable electrical equipment
- ▶ secondary hazards (falls, trips, noise and vibration).

Electric shock

- ▶ The effect of an electric shock can vary from a slight tingling sensation to death.
- ▶ The severity of the shock depends on the size of the current and voltage.
- ▶ Low-voltage shocks, below 110 V, are seldom fatal.
- ▶ Electric shock can also produce burns.

Battery-powered electrical tools will not give the user an electric shock.

Treatment of electric shock:

- ▶ Raise the alarm
- ▶ Switch off the power – if not possible, use insulating material to move victim from contact with the power supply
- ▶ Call for an ambulance
- ▶ If breathing, place victim in recovery position
- ▶ If not breathing, apply mouth-to-mouth resuscitation
- ▶ Treat any burns
- ▶ If person regains consciousness, treat for normal shock
- ▶ Remain with person until help arrives.

If high voltages are involved, inform police and electrical supply company but do not approach the victim.

Electrical fires and explosions

Twenty-five per cent of all fires have an electrical origin.

Causes:

▶ short circuits
▶ overheating of cables and equipment
▶ ignition of flammable gases and vapours
▶ ignition of combustible substances by static electrical discharges
▶ electric arcing
▶ static electricity – lightning strikes.

Portable electrical equipment

Twenty-five per cent of all electrical accidents are caused by portable electrical equipment.

Hazards:

▶ faulty cables, extension leads, plugs and sockets
▶ inadequate maintenance
▶ use in flammable or damp atmospheres
▶ misuse of equipment.

The safest position for an extension cable while in use is above head height.

Sources of high risks associated with the use of electricity

▶ working with poorly maintained electrical equipment;
▶ using electrical equipment in adverse or hazardous environments such as wet, flammable or explosive atmospheres;
▶ working on mains electricity supplies;
▶ contact with underground cables during excavation work; and
▶ contact with live overhead power lines.

Control measures

The main control measures are given in the Electricity at Work Regulations, which cover the following topics:

▶ the design, construction and maintenance of electrical systems, work activities and protective equipment;

▶ the strength and capability of electrical equipment;

▶ the protection of equipment against adverse and hazardous environments;

▶ the insulation, protection and placing of electrical conductors;

▶ the earthing of conductors and other suitable precautions;

▶ the integrity of referenced conductors;

▶ the suitability of joints and connections used in electrical systems;

▶ means for protection from excess current;

▶ means for cutting off the supply and for isolation;

▶ the precautions to be taken for work on equipment made dead;

▶ working on or near live conductors;

▶ adequate working space, access and lighting; and

▶ the competence requirements for persons working on electrical equipment to prevent danger and injury.

Specific controls for electrical hazards

▶ A management system that ensures safe installation, operation and maintenance

▶ Training at all levels – induction, supervisory and technical

▶ Safe systems of work include permits to work on live electricity

▶ Selection of suitable equipment for the work and environment by considering the following:

▷ atmosphere (flammable, damp)

▷ weather conditions

▷ high or low temperatures

▷ dirty or corrosive conditions

155

 ▷ standard of installation
 ▷ rated operating conditions
► The use of protective systems
► Inspection and maintenance strategies.

Protective systems:

► Fuse (and circuit breaker)
► Insulation
► Isolation
► Reduced low-voltage systems
► Residual current devices
► Double insulation.

Protection against contact with live buried power lines

► Check for any obvious signs of underground services, for example valve covers or patching of the road surface.
► Ensure that the excavation supervisor has the necessary service plans and is competent to use them to locate underground services.
► Ensure that all excavation workers are trained in safe digging practices and emergency procedures.
► Use locators to trace any services and mark the ground accordingly. A series of trial holes should be dug by hand to confirm the position of the pipes or cables. This is particularly important in the case of plastic pipes, which cannot be detected by normal locating equipment.
► In areas where underground services may be present, only hand digging should be used with insulated tools. Spades and shovels should be used rather than picks and forks, which are more likely to pierce cables.
► Assume that all cables are 'live' unless it is known otherwise.
► Hand-held power tools should not be used within 0.5 m of the marked position of an electricity cable. Collars should be fitted to the tools so that initial penetration of the surface is restricted.

156

▶ Any suspected damage to cables must be reported to the service providers and the health and safety enforcement authority.

▶ All exposed cables should be backfilled with fine material such as dry sand or small gravel.

▶ The service plans must be updated when the new cables have been laid.

Emergency procedures

Following an electrical incident, the emergency procedures are:

▶ The isolation of the electrical device that has caused the emergency by switching off and disconnecting the supply.

▶ If electric shock or fire is the main emergency, appropriate procedures should be activated.

▶ If there is a fire as a result of electrical equipment malfunction, the fire procedures should be activated.

▶ If the emergency could affect hazardous equipment or processes, emergency plant shutdown and isolation should be put into action.

▶ Work must not resume after an emergency until a competent person has checked and declared that all electrical equipment affected by the emergency is safe.

▶ Following the incident, an investigation should take place, a report produced and, if appropriate, a RIDDOR report submitted to the HSE.

Inspection and maintenance strategies

The particular areas of interest for inspection and maintenance are:

▶ the competence of the inspection and maintenance team

▶ the cleanliness of insulator and conductor surfaces

▶ the mechanical and electrical integrity of all joints and connections

▶ the integrity of mechanical mechanisms, such as switches and relays

▶ the calibration, condition and operation of all protection equipment, such as circuit breakers, RCDs and switches

- ▶ isolation procedures
- ▶ inspection and maintenance records
- ▶ permits to work for work on live electricity.

Inspection strategies

Any strategy for the inspection of electrical equipment, particularly portable appliances, should involve the following considerations:

- ▶ a means of identifying the equipment to be tested
- ▶ the number and type of appliances to be tested
- ▶ the competence of those who will undertake the testing (whether in-house or bought-in)
- ▶ the legal requirements for portable appliance testing (PAT) and other electrical equipment testing and the guidance available
- ▶ organisational duties of those with responsibilities for PAT and other electrical equipment testing
- ▶ test equipment selection and recalibration
- ▶ the development of a recording, monitoring and review system
- ▶ the development of any training requirements resulting from the test programme.

Portable electrical appliance testing

General points:

- ▶ user checks – at least once a week
- ▶ formal visual inspection – by a trained person
- ▶ combined testing and inspection – by a competent person using testing equipment
- ▶ the frequency of inspection and testing depends on usage rate and conditions – normally determined by a risk assessment
- ▶ records of inspection and testing required.

Advantages of portable appliance testing	Limitations of portable appliance testing
▶ early indication of serious faults	▶ too frequent testing of fixed equipment can lead to excessive costs
▶ discovery of incorrect equipment or electrical supply	▶ some unauthorised equipment never tested because of lack of records
▶ incorrect fuses found	▶ equipment misused or overused between tests
▶ reduction in the number of electrical accidents	
▶ monitoring misuse of portable appliances	▶ the inclusion of trivial faults leads to long lists and the overlooking of significant faults
▶ check on selection procedures	▶ the level of competence of tester
▶ increase awareness of hazards	▶ the use of incorrectly calibrated test equipment
▶ introduction of a more regular maintenance regime	

Control measures for working near overhead power lines

No vehicles, plant or equipment should be brought closer than:

▶ 15 m of overhead lines suspended from steel towers
▶ 9 m of overhead lines supported on wooden poles.

If a closer approach required, lines should be made dead and a permit to work used.

If vehicles need to move beneath power lines:

▶ the roadway should be covered by goalposts covered with warning tape;
▶ bunting should be suspended, level with the top of the goalposts and just above ground level between poles across the site along the length of the power line;
▶ suitable warning signs should be placed on either side of the roadway on each side of the power line to warn of the overhead power line; and
▶ reach of vehicles/equipment should be restricted.

159

At the **tender stage**, the existence of any overhead lines across the site must be noted and contact made with both the power supply company and the local authority.

For work at or near overhead power lines, the following **hierarchy of risk control** should be considered:

1 avoid working under or near overhead lines; if this is not possible,
2 divert all overhead lines clear of the work area or, if it is not feasible,
3 make lines dead while the work is in progress or, if this is not feasible,
4 work around the live overhead lines using the precautions outlined below.

Precautions

The precautions depend on the nature of the work at the site but the following recommendations should be possible for most work:

▶ Clearance – the safe clearance required beneath the overhead lines should be ascertained from the owner of the line;
▶ Exclusion – plant, equipment or hand tools that could reach beyond the safe clearance limit should not be taken under the line;
▶ Modifications – plant such as cranes and excavators should be modified by the addition of suitable physical restraints so that it cannot reach beyond the safe clearance limit;
▶ Additions – cranes with telescopic or fly jibs may need additional restraining devices to prevent alteration in length of jib or angle of fly jib; and
▶ Supervision – access for plant and materials should be under the direct supervision of a suitable person.

Fire safety

Learning outcomes	
Describe the principles of fire initiation, classification and spread and the additional fire risks caused by construction activities in an existing workplace	☐
Outline the principles of fire risk assessment	☐
Outline the principles of fire prevention and the prevention of fire spread in construction workplaces	☐
Identify the appropriate fire detection, fire alarm systems and fire-fighting equipment for construction activities	☐
Outline the requirements for an adequate and properly maintained means of escape in the construction workplace	☐
Outline the factors which should be considered when implementing a successful evacuation of a construction workplace in the event of a fire	☐

⚷ Key revision points

The fire triangle	☐
The legal framework for the control of fire hazards	☐
The basic principles of fire prevention – classification, methods of extinction and principles of heat transmission	☐
The common causes of fire, its spread and consequences	☐
Additional fire risks within the construction industry and the consequences of fires during construction activities	☐
Key elements of a fire risk assessment	☐
The principles of fire detection, protection, warning and means of escape	☐
The provision, maintenance and testing of fire-fighting equipment	☐
Emergency planning and training	☐

Principles of fire initiation, classification and spread and fire risks caused by construction activities

Basic principles

The fire triangle consists of:

▶ Fuel – solids, liquids and gases
▶ Ignition source
▶ Oxygen – from air, oxidising materials.

Classes of fire

▶ A – carboniferous solids
▶ B1 – liquids soluble in water
▶ B2 – liquids insoluble in water

162

▶ C – gases
▶ D – metals
▶ F – high temperature cooking oils or fats.
▶ Electrical fires do not constitute a separate fire class.

Sources of ignition

▶ Naked flames
▶ External sparks
▶ Internal sparking
▶ Hot surfaces
▶ Static electricity.

Control of ignition sources

The controls are:

▶ Replace a potential source with a safer alternative (e.g. radiant heaters with fixed convector heaters).
▶ Separate combustible materials.
▶ Minimise the use of portable heating appliances.
▶ Replace hot work with alternative methods such as cold cutting.
▶ Operate permit-to-work schemes for hot work processes and undertake regular checks.
▶ Operate a safe smoking policy.
▶ Control, inspect and monitor ignition hazards (e.g. overloading of power supplies).
▶ Ensure electrical, mechanical and gas equipment is installed, used, maintained and protected in accordance with the manufacturer's instructions.
▶ Turn off equipment when it is not attended or being used.
▶ Protect storage areas containing gas cylinders and other flammable materials from arson or vandalism.
▶ Do not permit bonfires on site.

Methods of extinction

▶ Cooling – reducing ignition temperature
▶ Smothering – limiting oxygen available

▶ Starving – limiting fuel supply
▶ Chemical reaction – interrupting the chain of combustion.

Principles of heat transmission and fire spread

▶ Convection – through air movements (floor to floor)
▶ Conduction – through conducting materials (girders)
▶ Radiation – emission of heat waves (bar heaters)
▶ Direct burning – contact with flames (rubbish catching fire).

Causes of fire

▶ rubbish (particularly if burnt)
▶ electrical short circuits
▶ portable heaters
▶ friction causing sparks (hand grinders)
▶ oil heating plant
▶ cooking appliances
▶ arson (often by children)
▶ hot surfaces (welding)
▶ cigarette smoking.

Fire spreads via:

▶ gases/sprays
▶ foam plastics
▶ packaging
▶ previous spillage
▶ flammable dusts.

Fire spreads due to:

▶ delayed discovery
▶ alarm not raised
▶ fire service not informed
▶ lack of barriers
▶ fire procedures not known
▶ fire extinguishers unavailable, not serviced or not used
▶ air flow – particularly if fire doors not closed
▶ lack of or poor emergency lighting

164

▶ flammable materials
▶ flammable dusts.

Additional risk of fire in construction

▶ final fire protection of the building not in place;
▶ timber frames are more vulnerable to fire before the external finishes are in place;
▶ many construction work processes, such as welding, heating bitumen for roofing, cutting and grinding metals and concrete;
▶ enormous quantities of packaging materials; and
▶ sub-contractors occupying the building at the same time, some of them extremely temporary, for only a few days.

Fire risk assessment

Required by the Regulatory Reform (Fire Safety) Order 2005.

Stages of the fire risk assessment

1 Identify the fire hazards – identify sources of heat/ignition, unsafe acts (smoking) and unsafe conditions (storage of combustible materials).
2 Identify all locations and persons at risk.
3 Reduce the risks by the introduction of controls.
4 Record the findings.
5 Monitor and review.

Hazards:

▶ sources of fuel
▶ oxygen depletion
▶ flames and heat
▶ smoke
▶ gaseous combustion products
▶ structural failure of the building
▶ combustible materials on construction sites.

165

Sources of heat or ignition on construction sites:

- ▶ refuelling of vehicles
- ▶ dismantling fuel tank structures
- ▶ buried gas and electrical services
- ▶ bonfires
- ▶ arson.

Identifying sources of fuel on construction sites:

- ▶ building materials (e.g. composite panels and timber)
- ▶ waste material
- ▶ flammable liquids and substances (e.g. paints and thinners)
- ▶ scaffold sheets and boards
- ▶ fuel for vehicles and heaters
- ▶ acetylene
- ▶ packaging materials.

Other factors:

- ▶ time to detect and warn
- ▶ safe egress
- ▶ adequacy of fire signs
- ▶ provision of fire-fighting equipment
- ▶ fire training – fire drills, appliance use and emergency teams
- ▶ maintenance of fire equipment
- ▶ consultation with employees.

Temporary workplaces

Construction sites, temporary classrooms and buildings, exhibitions, festivals and fetes.

The requirements depend on:

- ▶ the size of the premises;
- ▶ the number of persons working or visiting the site at any one time;
- ▶ the degree of flammability of the materials used to construct the workplace or being used in the workplace;
- ▶ height above or below the ground floor and distance to a place of safety;

▶ the remoteness of the workplace relative to water supplies and/
or fire and rescue services.

Temporary accommodation units:

▶ erect away from the construction site;
▶ include in emergency planning;
▶ careful installation and use of heaters and cooking equipment;
▶ competent installation and maintenance of electrical services; and
▶ the installation of litter skirts.

Fire prevention and prevention of fire spread

Control measures

▶ elimination or reduction in the use and storage of flammable
materials
▶ electrical safety
▶ control of ignition sources
▶ effective systems of work
▶ safe refuelling procedures for vehicles
▶ safe heating and cooking equipment and procedures
▶ good housekeeping
▶ storage facilities for small quantities (50 litres) of highly
flammable or flammable liquids.

Fire detection, fire alarm and fire-fighting equipment

Fire extinguishers

Fire extinguishers are coloured red over 95% of surface.

Colour (5%)	Contents	Class/type of fire
Red	Water	A
Cream	AFFF	A (possibly B)
Cream	Foam	B and C
Black	Carbon dioxide	B and electrical
Blue	Dry powder	B, C and electrical

Other fire-fighting equipment

▶ fire blankets
▶ hose reels
▶ sprinklers.

Emergency lighting

▶ Regular testing required.

Siting, maintenance and testing of fire-fighting equipment

▶ sited in areas of significant risk (at least one extinguisher per 200 m² of floor space with a minimum of one per floor);
▶ arrangements for hot working;
▶ annual maintenance check of fire extinguishers; and
▶ weekly check of fire alarms and emergency lighting.

Fire training in the use of the equipment for all staff is essential.

Fire detection and warning

▶ fire notices
▶ alarms – such as a gong, whistle or air horn and smoke/heat detectors
▶ audibility throughout premises
▶ regularly tested
▶ records of all tests
▶ vigilance of workers following training.

Automatic fire detection may be needed if:

▶ there are areas where people are isolated or remote and could become trapped;
▶ sleeping accommodation is provided;
▶ there are areas where a fire can develop unobserved (e.g. a warehouse).

Means of escape

▶ Doors – clear of obstructions and opening in direction of escape
▶ Escape routes (stairways and passageways) directly to a place of safety
▶ Lifts should not be used
▶ Lighting, including emergency lighting
▶ Directional and other signs
▶ Escape times
▶ Assembly points
▶ Building plans to include record of emergency escape plans.

Structural design measures

▶ Fire loading of the building
▶ Surface spread of fire
▶ Fire resistance of structural elements
▶ Insulating materials
▶ Fire compartmentation.

Evacuation of a construction workplace

Emergency planning and training

▶ evacuation procedures – fire notices
▶ appointment of fire marshals
▶ fire drills
▶ the location of fire points and assembly points
▶ roll-call
▶ provision for the infirm and disabled
▶ accident reporting and investigation
▶ rescue from excavations and confined spaces.

169

Fire procedures for people with a disability

Emergency planning needs to consider the special requirements of people with physical or mental disabilities. Such issues include:

▶ the identification of those who may need special help to evacuate the building;

▶ the allocation of responsibility to specific staff to identify and assist people with a disability during emergencies;

▶ plan suitable escape routes;

▶ have a system in place that enables a person with a disability to summon help in emergencies; and

▶ include the arrangements for disabled persons in all staff fire training programmes.

General points

▶ Emergency procedures required for fire, explosions, flooding or structural collapse.

▶ A hot work permit should include the need for a fire extinguisher and checks for signs of fire on completion of the work.

▶ A hot work permit would be needed when cutting steel with an angle grinder and soldering pipework in a central heating system.

▶ Frost around the valve on a liquid petroleum gas cylinder indicates that the valve is leaking.

▶ Cylinders containing liquid petroleum gas should be stored in a locked external compound at least 3 m from any oxygen cylinders.

▶ The minimum distance that oxygen cylinders should be stored away from propane, butane and other inflammable gases is 3 m.

▶ Butane and propane are heavier than air and therefore represent a ground level fire hazard.

2.8

Chemical and biological health hazards and risk control

Learning outcomes	
Outline the forms of, the classification of, and the health risks from exposure to hazardous substances	☐
Explain the factors to be considered when undertaking an assessment of the health risks from substances commonly encountered in construction workplaces	☐
Explain the use and limitations of workplace exposure limits (WELs), including the purpose of long-term and short-term exposure limits	☐
Outline control measures that should be used to reduce the risk of ill-health from exposure to hazardous substances	☐
Outline the hazards, risks and controls associated with specific agents	☐
Outline the basic requirements related to the safe handling and storage of waste on construction sites	☐

🔖 Key revision points

The forms and classification of hazardous chemical and biological agents ☐

The functions of various human body organs and their vulnerability to hazardous substances ☐

The difference between acute and chronic effects ☐

The requirements of the COSHH Regulations – assessment and recommended control measures ☐

The definitions and limitations of workplace exposure limits (WELs) ☐

The principles of prevention and good practice ☐

The hierarchy of controls for hazardous substances ☐

The use of engineering controls (LEV) and personal protective equipment ☐

The health risks and controls associated with asbestos and other specific agents ☐

Health surveillance ☐

Emergency and maintenance controls ☐

The safe handling and storage of waste ☐

Forms and classification of, and health risks from exposure to, hazardous substances

Agents

Chemical agents	Biological agents
▶ Dusts – including respirable dust (fine dust that remains in the lungs)	▶ Fungi
	▶ Moulds
▶ Fibres	
▶ Gases	▶ Bacteria
▶ Vapours – substances very close to their boiling points (e.g. solvents)	▶ Viruses
▶ Liquids	
▶ Mists – similar to vapours but closer to the liquid phase (e.g. paint sprays)	
▶ Fume – condensed metallic particles (e.g. welding fume)	

GHS classification of hazardous substances

Globally Harmonized System (GHS) classifications have replaced CHIP 4 ones.

▶ The **health hazard** classification combines the CHIP 4 '**harmful**' and '**irritant**' classifications – involves limited health risks and/ or repeated contact with skin can cause a sensitised or allergic inflammation.

▶ The **serious health hazard** classification replaces CHIP 4 '**toxic**' – produces serious health risks.

▶ The **corrosive** classification is very similar to the CHIP 4 '**corrosive**'– may destroy living tissue.

▶ The **acute toxicity** classification is similar to the '**very toxic**' CHIP 4 – very hazardous to health when inhaled, swallowed or in contact with the skin and could be fatal.

▶ Carcinogenic – may induce cancer.

▶ Mutagenic – may induce hereditary genetic defects.

The UN Globally Harmonized System (GHS) on the classification and labelling of chemicals has been adopted by the EU as the CLP Regulation.

173

Under the CLP Regulation there are:

▶ new scientific criteria to assess hazardous properties of chemicals;
▶ two new harmonised hazard warning symbols for labels (known as 'pictograms');
▶ a new design for existing symbols;
▶ new harmonised hazard warning (H) and precautionary statements (P) for labels; and
▶ two new signal words – 'Danger' and 'Warning'.

A hazard statement describes the nature of the hazard in the substance or mixture.

A precautionary statement describes recommended measure(s) to minimise or prevent adverse effects resulting from exposure to a hazardous substance.

If the substance has a more severe hazard, the label includes the signal word 'Danger'; in case of less severe hazards, the signal word is 'Warning'.

Acute effects are of short duration, normally reversible and appear shortly after exposure to a hazardous substance (e.g. asthma attacks).

Chronic effects develop over a period of time after repeated exposure to a hazardous substance and are often irreversible.

Assessment of health risks

Health hazards associated with construction activities

▶ noise
▶ vibration
▶ dust (including asbestos)
▶ cement dust and wet cement
▶ solvents, cleaners and paints
▶ tetanus and other biological agents
▶ manual handling.

174

Categories of health risks	Types of health risk
▶ Chemical – exhaust fumes, paint solvents, asbestos ▶ Biological – legionella, other pathogens, hepatitis ▶ Physical – noise, vibration, radiation ▶ Psychological – stress, violence ▶ (Ergonomic – musculoskeletal disorders)	▶ skin contact with irritant substances, leading to dermatitis; ▶ inhalation of respiratory sensitisers, triggering immune responses such as asthma; ▶ badly designed workstations requiring awkward body postures or repetitive movements, resulting in upper limb disorders, repetitive strain injury and other musculoskeletal conditions; ▶ noise levels which are too high, causing deafness and conditions such as tinnitus; ▶ too much vibration, for example from hand-held tools leading to hand–arm vibration syndrome and circulatory problems; ▶ exposure to ionising and non-ionising radiation, including ultraviolet in the sun's rays, causing burns, sickness and skin cancer; ▶ infections ranging from minor sickness to life-threatening conditions, caused by inhaling or being contaminated with microbiological organisms; and ▶ stress causing mental and physical disorders.

Routes of entry into the body

▶ Inhalation via the lungs
▶ Absorption via the skin
▶ Ingestion via the stomach
▶ Injection through the skin.

Major human body systems

There are five major functional systems within the body. These are listed together with typical diseases and related hazardous substances:

Body system	Illness or disease	Hazardous substance
Respiratory	bronchitis, asthma, fibrosis, cancer	dusts, asbestos
Nervous	anxiety, epilepsy, loss of consciousness	solvents, lead
Cardiovascular	headaches, loss of consciousness	benzene, carbon monoxide
Urinary	cirrhosis, cancer	chlorinated hydrocarbons, mercury
Skin	irritant contact dermatitis and allergic contact dermatitis	detergents, turpentine, epoxy resin

Control of Substances Hazardous to Health Regulations

Requirements of COSHH

▶ Suitable and sufficient health risk assessment
▶ Adequate control of exposure of employees
▶ Proper use of control measures provided
▶ Maintain control measures
▶ Monitor employees exposed to hazardous substances (Schedule 5)
▶ Health surveillance (Schedule 6)
▶ Information, instruction and training.

COSHH assessment

1 Identify hazardous substances
2 Gather information
3 Evaluate the risks
4 Decide on control measures
5 Record the assessment
6 Review the assessment.

Sources of information

▶ Product labels
▶ Safety data sheets

176

▶ HSE Guidance Note EH 40
▶ Trade association publications
▶ Industrial codes of practice
▶ Specialist reference manuals.

Survey techniques for health risks

Include:

▶ Stain tube detector. The advantages of this technique are
 that it is quick, relatively simple to use and inexpensive. The
 disadvantages are:
 ▷ it cannot be used to measure concentrations of dust or fume
 ▷ the accuracy of the reading is approximately ± 25%
 ▷ it will yield false readings if other contaminants present react
 with the crystals
 ▷ it can only give an instantaneous reading, not an average
 reading over the working period (TWA)
 ▷ the tubes are very fragile with a limited shelf life
▶ Passive sampling, including the static dust sampler
▶ Sampling pumps and heads
▶ Direct reading instruments (MIRA)
▶ Vane anemometers and hygrometers
▶ Smoke tube and dust observation lamp.

Workplace exposure limits (WELs)

Workplace exposure limits (WELs) must not be exceeded.

Two categories (similar to previous MEL and OES):

| Category 1 | carcinogenic and mutagenic substances – exposure must be reduced as low as is reasonably practicable below the WEL |
| Category 2 | all other hazardous substances – exposure controlled by principles of good practice |

There are two time-weighted averages (TWA):

▶ the long-term exposure limit (LTEL) – over 8 hours
▶ the short-term exposure limit (STEL) – 15 minute reference
 period.

177

Limitations of exposure limits

▶ They are specifically quoted for an 8-hour period (with an additional STEL for many hazardous substances). Adjustments must be made when exposure occurs over a continuous period longer than 8 hours.

▶ They can only be used for exposure in a workplace and not to evaluate or control non-occupational exposure (e.g. to evaluate exposure levels in a neighbourhood close to the workplace, such as a playground).

▶ WELs are only approved where the atmospheric pressure varies from 900 to 1100 millibars. This could exclude their use in mining and tunnelling operations.

▶ They should also not be used when there is a rapid build-up of a hazardous substance due to a serious accident or other emergency.

▶ The fact that a substance has not been allocated a WEL does not mean that it is safe.

Control measures

The general principles of prevention are:

1 the avoidance of risks;
2 the evaluation of unavoidable risks;
3 controlling hazards at source;
4 replacing the dangerous by less or non-dangerous alternatives;
5 adapting work to the individual;
6 adapting to technical progress;
7 developing a coherent prevention policy;
8 giving priority to collective over individual protective measures; and
9 giving appropriate instructions to employees.

Principles of Good Practice for the control of exposure to substances hazardous to health

1 Design and operate processes and activities to minimise emission, release and spread of substances hazardous to health.

2 Take into account all relevant routes of exposure – inhalation, skin absorption and ingestion – when developing control measures.

3 Control exposure by measures that are proportionate to the health risk.

4 Choose the most effective and reliable control options which minimise the escape and spread of substances hazardous to health.

5 Where adequate control of exposure cannot be achieved by other means, provide, in combination with other control measures, suitable PPE.

6 Check and review regularly all elements of control measures for their continuing effectiveness.

7 Inform and train all employees on the hazards and risks from the substances with which they work and the use of control measures developed to minimise the risks.

8 Ensure that the introduction of control measures does not increase the overall risk to health and safety.

Hierarchy of control measures

1 eliminate the risks by either designing them out or changing the process;

2 substitution of hazardous substances with others which are less hazardous;

3 isolation, using enclosures, barriers or worker segregation;

4 engineering controls such as guarding, the provision of local exhaust ventilation systems, the use of reduced voltage systems or residual current devices;

5 management controls such as safe systems of work, training, job rotation and supervision; and

6 finally, as a last resort, the provision of personal protective equipment such as ear defenders or respiratory protective equipment.

Hence, measures for preventing or controlling exposure to hazardous substances include one or a combination of the following:

▶ elimination of the substance
▶ substitution of the substance (or reduction in the quantity used)
▶ total or partial enclosure of the process

- local exhaust ventilation
- dilute or general ventilation
- reduction of the number of employees exposed to a strict minimum
- reduced time exposure by task rotation and the provision of adequate breaks
- good housekeeping
- training and information on the risks involved
- effective supervision to ensure that the control measures are being followed
- personal protective equipment (such as clothing, gloves and masks)
- welfare (including first-aid)
- medical records
- health surveillance.

Preventative control measures

Engineering controls

These include:

- segregation of people from the process (fume cupboard)
- modification of the process to reduce human contact with the hazardous substances
- local exhaust ventilation comprising:
 - ▷ a collection hood and intake
 - ▷ ventilation ducting
 - ▷ a filter or other air cleaning device
 - ▷ a fan
 - ▷ an exhaust duct
- dilution or general ventilation.

Supervisory or people controls

Additional supervisory controls when hazardous substances are involved are:

- reduced time exposure – thus ensuring that workers have breaks in their exposure periods;

▶ reduced number of workers exposed – only persons essential to the process should be allowed in the vicinity of the hazardous substance; walkways and other traffic routes should avoid any area where hazardous substances are in use;

▶ eating, drinking and smoking must be prohibited in areas where hazardous substances are in use;

▶ any special rules, such as the use of personal protective equipment, must be strictly enforced.

Personal protective equipment

Personal protective equipment (PPE) is the control measure of **last resort**. It is covered by the Personal Protective Equipment at Work Regulations, which require:

▶ PPE to be suitable for wearer and task;

▶ PPE to be compatible with any other personal protective equipment;

▶ a risk assessment to be undertaken to determine best PPE;

▶ a suitable PPE maintenance programme;

▶ suitable storage arrangements for the PPE;

▶ information, instruction and training for the user of the PPE;

▶ supervision of the use of PPE; and

▶ a system for reporting defects.

Notes:

1 The Regulations do not cover respiratory protective equipment (RPE).

2 The risk assessment and method statement for the particular construction activity should include information on personal protective equipment.

Types of personal protective equipment	Types of respiratory protective equipment
▶ hand and skin protection	▶ filtering half-mask
▶ eye protection	▶ half-mask respirator
▶ protective clothing – safety gloves and boots, hard hats, aprons	▶ full-face mask respirator
	▶ powered respirator

181

Health surveillance

Health surveillance is important for:

- ▶ detecting ill-health effects at an early stage;
- ▶ providing data to help employers evaluate health risks;
- ▶ enabling employees to raise concerns;
- ▶ highlighting lapses in workplace control measures; and
- ▶ providing an opportunity to reinforce training and education of employees

Health surveillance is not the same as health monitoring. Types of health surveillance include:

- ▶ checking individuals on a regular basis;
- ▶ checking individuals as a result of statistical evidence of a potential problem; and
- ▶ when a substance listed in Schedule 6 of COSHH is being used. In these cases, health surveillance should take place at intervals not exceeding 12 months and records should be kept for 40 years.

Maintenance and emergency controls

Maintenance involves:

- ▶ the cleaning, testing and, possibly, the dismantling of equipment;
- ▶ the changing of filters in extraction plant or entering confined spaces;
- ▶ the handling of hazardous substances and the safe disposal of waste material;
- ▶ a permit-to-work procedure may also need to be in place, since the control equipment will be inoperative during the maintenance operations;
- ▶ the keeping of records of maintenance for at least 5 years.

The following points should be considered when **emergency procedures** are being developed:

- ▶ the possible results of a loss of control (e.g. lack of ventilation);
- ▶ dealing with spillages and leakages (availability of effective absorbent materials);
- ▶ raising the alarm for more serious emergencies;

▶ evacuation procedures, including the alerting of neighbours;
▶ fire-fighting procedures and organisation;
▶ availability of respiratory protection equipment; and
▶ information and training.

Specific agents

Asbestos

A fine, sharp and hard fibrous dust of respirable dust size which can become lodged in the lungs, causing damage to the lining of the lungs over a period of many years. Asbestos is a possible chronic health risk. It can lead to one of the following diseases:

▶ asbestosis or fibrosis (scarring) of the lungs;
▶ diffuse pleural thickening;
▶ lung cancer;
▶ mesothelioma – cancer of the lining of the lung.

Duty to manage asbestos

The Control of Asbestos at Work Regulations require:

▶ the determination of the location and condition of materials likely to contain asbestos (ACMs)
▶ the presumption that materials contain asbestos unless there is strong counter-evidence
▶ up-to-date records of the location and condition of the ACMs or presumed ACMs
▶ an assessment of the risk of anyone being exposed to fibres from ACMs
▶ a plan to manage the risks from ACMs
▶ the necessary steps to action the plan
▶ the periodic review and monitoring of the plan
▶ provision of information and asbestos awareness training to anyone who is liable to work on ACMs or otherwise disturb them.

In addition, the Regulations include the following main provisions:

1 A single tighter control limit for all types of asbestos;

2 Specific training requirements for those working with asbestos;
3 A clear hierarchy of controls to be used to reduce exposure.

Types of asbestos survey

▶ The management survey – the location and assessment of condition of ACMs, used to compile the asbestos register.
▶ The refurbishment or demolition survey – most invasive and detailed and used to locate and describe, as far as reasonably practicable, all ACMs in the area where the refurbishment work will take place or in the whole building if demolition is planned. Refurbishment and demolition surveys should only be conducted in unoccupied areas to minimise risks to the public or employees on the premises.

Stages in the control of asbestos

▶ **Identification** of the presence of asbestos to compile the asbestos register
▶ **Assessment** is an evaluation as to whether the location or the condition could lead to the asbestos being disturbed
▶ **Removal** must only be done by a licensed contractor
▶ **Control measures** during the removal of asbestos to ensure adequate protection of workers and others
▶ **Medical surveillance** in the form of a regular medical examination should be given to any employee who has been exposed to asbestos at levels above the action level.
▶ **Awareness training** is given to those employees:
 ▷ who are or who are liable to be exposed to asbestos or who supervise such employees
 ▷ who carry out work in connection with the employer's duties under these Regulations, so that they can carry out that work effectively
▶ Disposal of asbestos waste to an authorised asbestos waste site.

If asbestos is discovered during the performance of a contract, work should cease immediately and the employer should be informed.

Health hazards of other specific agents

Each agent is followed by a brief description and two possible illnesses.

Organic solvents – sensitisers and irritants that can cause dermatitis and liver failure. The minimum PPE requirements are impermeable overalls, apron, footwear, long gloves or gauntlets and chemically resistant goggles or visor. Respiratory protective equipment is also required if it cannot be demonstrated that exposure is below the appropriate workplace exposure limit (WEL). Organic solvents are often used with painting work.

Carbon dioxide – odourless gas, can cause headaches and loss of consciousness.

Nitrogen – odourless gas, in liquid form used as a freezing agent, can cause frostbite.

Isocyanates – sensitisers and irritants, can cause bronchitis and extreme 'asthma attack'. They are often used in paint spraying operations and require good ventilation. Respiratory Protective Equipment (RPE) should be worn.

Lead – dust can cause headaches, nausea and anaemia. Lead is used in roofing and plumbing work.

Carbon monoxide – odourless gas, can cause headaches and unconsciousness. Exposure may be prevented by ensuring that any work carried out in relation to gas appliances in domestic or commercial premises is undertaken by a competent Gas Safe Registered engineer.

Silica dust produced in masonry work can cause fibrosis and silicosis. Prevention is best achieved by the use of good dust extraction systems and respiratory protective equipment.

Cement dust – irritant and allergenic, can cause dermatitis, inflammation and irritation of the eye, irritation of the nose and throat and, possibly, chronic lung problems. PPE in the form of gloves, overalls with long sleeves, full-length trousers and waterproof boots must be worn.

Wet cement – irritant and allergenic, can cause dermatitis, serious burns or ulcers.

Wood dust can be hazardous, particularly from hard woods and composite boards (MDF), and can cause irritation to the eyes, nose and throat, leading to dermatitis, asthma, rhinitis and even cancer. An extraction system is essential and PPE in the form of gloves, respiratory protective equipment, overalls and eye protection.

Leptospira is a bacterium in rat urine that can cause anaemia and jaundice. Good, impervious protective clothing, particularly wellington boots, and the covering of any skin wounds are essential in these situations.

Legionella are bacteria in tepid water that can cause pneumonia and death. Where a plant is at risk of the development of legionella, the following is required:

▶ a written 'suitable and sufficient' risk assessment;
▶ the preparation and implementation of a written control scheme involving the treatment, cleaning and maintenance of the system;
▶ appointment of a named person with responsibility for the management of the control scheme;
▶ the monitoring of the system by a competent person; and
▶ record keeping and the review of procedures developed within the control scheme.

Note: Any potential source of legionella (e.g. cooling towers) must be given a special risk assessment by a competent person and records must be kept.

Hepatitis and other blood-borne viruses can cause nausea, jaundice and death. Possible control measures:

▶ prohibit eating, drinking and smoking in working areas where there is a risk of contamination;
▶ prevent puncture wounds, cuts and abrasions, especially in the presence of blood and body fluids;
▶ when possible avoid use of, or exposure to, sharps such as needles, glass or metal, or if unavoidable take care in handling and disposal;

186

▶ consider the use of devices incorporating safety features, such as safer needle devices and blunt-ended scissors;

▶ cover all breaks in exposed skin by using waterproof dressings and suitable gloves;

▶ protect the eyes and mouth by using a visor/goggles/safety spectacles and a mask, where splashing is possible;

▶ avoid contamination by using water-resistant protective clothing;

▶ wear rubber boots or plastic disposable overshoes when the floor or ground is likely to be contaminated;

▶ use good basic hygiene practices, such as hand washing;

▶ control contamination of surfaces by containment and using appropriate decontamination procedures; and

▶ dispose of contaminated waste safely.

Tetanus – a bacterium in soil enters a wound from an infected object and attacks the nerves that supply muscle tissue (lockjaw); can be fatal.

Hydrogen sulphide – poisonous gas with odour of rotten eggs (sewer gas). Causes irritation of the eyes, headaches and dizziness; may result in pulmonary oedema.

Generation and control of dust on a construction site

Cement dust – controlled by PPE in the form of gloves, overalls with long sleeves, full-length trousers and waterproof boots.

Wood dust – controlled by a good extraction system and PPE (gloves, respiratory protective equipment, overalls and eye protection).

Silica dust – controlled by a good dust extraction system and respiratory protective equipment.

Safe handling and storage of waste

Mainly regulated by:

▶ The Hazardous Waste Regulations – Special Waste Regulations still in Scotland

▶ The Environmental Protection Act (EPA) covers hazardous waste and spillage.

There are two categories of hazardous waste – absolute hazardous waste, such as fuel and oil, or mirror hazardous waste, such as sawdust.

The principal requirements of the EPA are as follows:

▶ to handle waste so as to prevent any unauthorised escape into the environment;
▶ to pass waste only to an authorised person;
▶ to ensure that a written description accompanies all waste; and
▶ to ensure that no person commits an offence under the Act.

There are several different categories of waste. The important ones are as follows:

▶ Controlled waste, comprising household, industrial or commercial waste
▶ Hazardous waste, which can only be disposed of using special arrangements (toxic, corrosive, carcinogenic or highly flammable).

The hazards associated with waste skips

include:

▶ being struck by vehicles;
▶ falling and slipping;
▶ failures of lifting equipment;
▶ striking overhead cables/obstructions;
▶ vehicle overturns; and
▶ runaway vehicles.

Requirements for waste skips

▶ sufficient strength to cope with its load;
▶ stability while being filled;
▶ a reasonably uniform load distribution within the skip at all times;
▶ the immediate removal of any damaged skip from service and the skip inspected after repair before it is used again;
▶ sufficient space around the skip to work safely at all times;

▶ the skip should be resting on firm level ground;

▶ the skip should never be overloaded or overfilled; and

▶ there must be sufficient headroom for the safe removal of the skip when it is filled.

The control measures for hazards associated with waste disposal

▶ All workers should wear suitable PPE, such as high-visibility jackets, gloves and suitable footwear.

▶ Training so that employees segregate hazardous and non-hazardous wastes on site and fully understand the risks and necessary safety precautions to be taken.

▶ The storage site should be protected against trespassers, fire and adverse weather conditions.

▶ If flammable or combustible wastes are being stored, adequate fire protection systems must be in place.

▶ Manual handling issues may need to be considered.

▶ All lifting equipment, including chains and shackles, must be subject to a periodic statutory examination.

▶ Separate storage required for incompatible waste streams.

▶ Finally, for liquid wastes, drains must be protected and bunds used to restrict spreading of the substance as a result of spills.

The four requirements for the transfer of waste from the site to final treatment or disposal are:

1 The waste must be clearly identified.
2 The waste must be securely stored with procedures to deal with spillages.
3 The waste must only be transferred to an authorised person.
4 Adequate records of waste consignments must be kept for at least 2/3 years.

Physical and psychological health hazards and risk control

Learning outcomes	
Outline the health effects associated with exposure to noise and appropriate control measures	☐
Outline the health effects associated with exposure to vibration and appropriate control measures	☐
Outline the health effects associated with ionising and non-ionising radiation and outline appropriate control measures	☐
Outline the causes and effects of stress at work and appropriate control measures	☐

🔑 **Key revision points**

The health effects of exposure to noise and vibration	☐
The legal exposure limits to noise and vibration	☐
The control measures for excessive vibration and noise	☐
The hazards and controls associated with ionising and non-ionising radiations	☐
The causes and prevention of workplace stress	☐

Noise

Health effects of noise

Human ear:

▶ Ear drum
▶ Cochlea
▶ Hair cells – damage irreversible
▶ Auditory nerve.

Acute effects:

▶ Temporary threshold shift
▶ Tinnitus
▶ Acute acoustic trauma.

Chronic effects:

▶ Noise-induced hearing loss
▶ Permanent threshold shift
▶ Tinnitus.

Noise assessments

Objective is to identify whether action levels have been reached.

Noise measurement

▶ Sound pressure level (SPL) – dB(A)

- Peak sound pressure
- Continuous equivalent noise level (L_{Eq}) – normally measured over an 8-hour period
- Daily personal exposure level ($L_{EP,d}$).

Requirements of Noise at Work Regulations

- assess noise levels and keep records;
- reduce the risks from noise exposure by engineering controls – only use hearing protection as a last resort;
- provide employees with information and training; and
- if a manufacturer of equipment, provide information on noise levels of the equipment.

Noise action levels:

- First action level – daily exposure level of 80 dB(A) – action advised
- Second action level – daily exposure level of 85 dB(A) – action obligatory
- Peak action level – 135 dB(C) and 137 dB(C)
- Exposure limit values – 87 dB(A) and 140 dB(C).

Noise control techniques include:

- Reduction of noise at source
- Change equipment or process
- Change speed
- Improve maintenance
- Relocate equipment
- Enclose equipment
- Screens or absorption walls
- Damping
- Lagging
- Silencers
- Isolation of workers
- Suitable warning signs.

Personal ear protection:

- Ear plugs
- Ear defenders (earmuffs).

Factors to be considered:

- Suitability (frequencies)
- Acceptability and comfort
- Durability
- Instruction in use
- Hygiene considerations
- Beards, hair and spectacles may reduce effectiveness of protection
- Maintenance and storage
- Cost.

Vibration

Ill-health due to vibration

Ill-health due to hand–arm vibration

- Hand–arm vibration syndrome (HAVS)
- Vibration White Finger (VWF)
- Carpal tunnel syndrome.

Ill-health due to whole-body vibration

- Severe back pain
- Reduced visual and manual control
- Increased heart rate and blood pressure
- Spinal damage.

Control of Vibration at Work Regulations

The requirements of the Regulations are to:

- assess the vibration risk to employees;
- decide if they are likely to be exposed above the daily exposure action value (EAV) or the daily exposure limit level (ELV) and if they are in either case, take appropriate action;
- provide information and training on health risks and controls to employees at risk;
- consult safety representatives or employee representatives about the risks and the plans to address these risks;

▶ keep a record of the risk assessment and control actions; and
▶ review and update the risk assessment regularly.

Exposure limit and action values

For hand–arm vibration (HAV):

▶ the daily exposure limit value normalised to an 8-hour reference period is 5 m/s^2
▶ the daily exposure action value normalised to an 8-hour reference period is 2.5 m/s^2.

For whole-body vibration (WBV):

▶ the daily exposure limit value normalised to an 8-hour reference period is 1.15 m/s^2
▶ the daily exposure action value normalised to an 8-hour reference period is 0.5 m/s^2.

Preventative and precautionary measures

The common measures used to control ergonomic ill-health effects are:

▶ identification of repetitive actions;
▶ the elimination of vibration by performing the job in a different way;
▶ undertake a risk assessment;
▶ ensure that the correct equipment (properly adjusted) is always used;
▶ introduce job rotation or frequent breaks so that workers have a reduced time exposure to the hazard;
▶ during the design of the job ensure that poor posture is avoided;
▶ issue employees with gloves and warm clothing;
▶ examine ill-health reports and absence records;
▶ introduce a programme of health surveillance;
▶ ensure that workers are given adequate information on the hazards and develop a suitable training programme;
▶ introduce a reporting system for employees to use so that concerns and any symptoms can be recorded and investigated;
▶ ensure that drill bits and other tools are kept sharp;

195

▶ ensure that a programme of preventative maintenance is introduced and include the regular inspection of items such as vibration isolation mountings; and

▶ keep up to date with advice from equipment manufacturers, trade associations and health and safety sources.

Additional information on whole-body vibration

The reasons for back pain in drivers include:

▶ poor posture while driving;
▶ incorrect adjustment of the driver's seat;
▶ sitting for long periods without being able to change position;
▶ difficulty in reaching all relevant controls due to poor design of the controls layout;
▶ frequent manual handling of loads;
▶ frequent climbing up and down from a high cab;
▶ excessive exposure to whole-body vibration.

Actions for controlling the risks from WBV need to ensure that:

▶ the driver's seat is correctly adjusted so that all controls can be reached easily and that the driver weight setting on the suspension seat, if available, is correctly adjusted. The seat should have a back rest with lumbar support;
▶ anti-fatigue mats are used if the operator has to stand for long periods;
▶ the speed of the vehicle is such that excessive jolting is avoided; speeding is one of the main causes of excessive WBV;
▶ all vehicle controls and attached equipment are operated smoothly;
▶ only established site roadways are used;
▶ only suitable vehicles and equipment are selected to undertake the work and cope with the ground conditions;
▶ the site roadway system is regularly maintained;
▶ all vehicles are regularly maintained, with particular attention being paid to tyre condition and pressures, vehicle suspension systems and the driver's seat;

196

▶ work schedules are regularly reviewed so that long periods of exposure on a given day are avoided and drivers have regular breaks;

▶ prolonged exposure to WBV is avoided for at-risk groups (older people, young people, people with a history of back problems and pregnant women); and

▶ employees are aware of the health risks from WBV, the results of the risk assessment and the ill-health reporting system. They should also be trained to drive in such a way that excessive vibration is reduced.

General points on vibration

▶ A person who is warm and dry is less likely to suffer from HAV.

▶ HAV can cause damage to blood vessels and nerves in hands and fingers.

▶ The effects of HAV can be reduced if the work is done in short intervals.

▶ Anti-vibration gloves may not always protect against vibration.

Radiation

Radiation hazards

Ionising radiation

Ionising radiation is produced by alpha and beta particles and gamma rays.

Harmful effects:

▶ Somatic – cell damage to the individual

▶ Genetic – cell damage to any future children of the individual

▶ Acute – nausea, vomiting, skin burns and blistering, collapse and death

▶ Chronic – anaemia; leukaemia and other types of cancer.

Sources

▶ Radon gas

▶ X-ray equipment

▶ Smoke detectors.

Non-ionising radiation

▶ Ultraviolet radiation – sun, arc welding
▶ Lasers – eye and skin burns, electricity etc.
▶ Infrared radiation – fires, furnaces, etc.
▶ Microwaves – cookers/ovens, mobile telephones.

Non-ionising radiation can harm the eyes and skin.

The HSE suggests the following **hierarchy of controls for outdoor working**:

▶ relocate some jobs inside a building or to a shady location;
▶ undertake some outdoor work earlier or later in the day;
▶ provide personal protection such as:
▶ wearing long sleeve shirts or loose clothing with a close weave;
▶ wearing hats with a wide brim;
▶ using a high factor sunscreen of at least SPF15 on any exposed skin;
▶ provide suitable education and training for outdoor workers;
▶ provide suitable information and supervision to instigate safe systems of work that protect workers from the sun.

Radiation protection strategies

Ionising radiation

Covered by the Ionising Radiations Regulations, which specify:

▶ Risk assessment
▶ Shielding
▶ Time (reduced time exposure)
▶ Distance
▶ Training
▶ Personal protective equipment
▶ No food or drink consumption near exposed areas
▶ Signs and information
▶ Medical surveillance
▶ Maintenance and inspection controls
▶ Emergency procedures
▶ Radiation protection supervisor – appointed to advise on compliance with Regulations

▶ Radiation protection adviser – appointed to advise the radiation protection supervisor.

Non-ionising radiation

▶ Eye protection
▶ Skin protection (gloves and/or creams)
▶ Fixed shields and non-reflective surfaces (screen must be in place before welding operations begin to protect others in the area from getting arc eye)
▶ Interlocking guards on microwaves.

Stress

Work-related stress is the reaction people have to excessive pressure or other types of demand placed on them.

It is the reaction of the body to excessive mental pressure that may lead to ill-health.

Symptoms of stress

There are four groups of stress symptoms:

▶ physical – chest pains and frequent colds;
▶ behavioural – smoke and drink more;
▶ emotional – irritable and depressed; and
▶ cognitive – unable to concentrate and very anxious.

Causes:

▶ The job itself – unrealistic targets, boring, repetitive, insufficient training
▶ Individual responsibility – ill-defined roles, too much responsibility, too little control to influence outcome
▶ Working conditions – lack of privacy or security, unsafe practices, threat of violence, excessive noise
▶ Management attitudes – negative health and safety culture, poor communication, lack of support in a crisis
▶ Relationships with colleagues – bullying, harassment.

Prevention:

- ▶ Take a positive attitude to stress
- ▶ Take employee concerns seriously
- ▶ Effective communication and consultation
- ▶ Develop a policy on stress
- ▶ Relevant training
- ▶ Employee appraisal system
- ▶ Discourage excessive hours at work
- ▶ Encourage lifestyle changes
- ▶ Monitor incidents of bullying, etc.
- ▶ Avoid blame culture
- ▶ Set up a confidential counselling advice service.

Working at height – hazards and risk control

Learning outcomes	
Explain the hazards of working at height and outline the precautions necessary to control them	☐
Explain safe working practices for access equipment and roof work	☐
Outline control measures necessary to protect other persons not involved in the work at height	☐
Outline control measures to reduce risk when working over or near to water	☐

🔑 **Key revision points**

Examples of working at height ☐

Hierarchy of controls when working at height ☐

Controls and precautions when working at height ☐

Emergency rescue plans ☐

Different types of access equipment used in construction (scaffolding, ladders, stepladders, trestles, staging platforms) ☐

The hazards and controls required for access equipment (including leading edge protection systems) ☐

Types of fall arrest equipment ☐

Hazards and precautions during roof work ☐

Protection of the public during construction work ☐

The issues involved with working over or near water ☐

Working at height hazards and controls

Work at Height Regulations

The Regulations have no minimum height requirement for work at height. They cover many activities, including:

▶ working on a scaffold or from a mobile elevating work platform (MEWP);

▶ sheeting a lorry;

▶ tree surgery;

▶ using cradles or rope for access to a building or other structure;

▶ working near an excavation area or cellar opening;

▶ working on staging or trestles;

▶ using a ladder/stepladder or kick stool for shelf filling, window cleaning and the like; and
▶ working in a mine shaft or chimney.

The Regulations give the following hierarchy of control:

1 Avoid working at height if possible
2 Use an existing safe place of work
3 Provide work equipment to prevent falls (e.g. guard rails)
4 Mitigate distance and consequences of a fall (e.g. air bags, safety nets and, as a last resort, safety harnesses)
5 Instruction, training and supervision.

Work at height can be reduced by completing some of the work at ground level.

A risk assessment for working at height should consider the following issues:

▶ the nature and duration of the work;
▶ the competence level of all those involved with the work and any additional training requirements;
▶ the required level of supervision;
▶ use of guard rails, toe boards, working platforms and means of access and egress;
▶ required personal protective equipment, such as helmets and harnesses;
▶ the presence of fall arrest systems, such as netting or soft landing systems;
▶ the health status of the workers;
▶ the possible weather conditions; and
▶ compliance with the relevant legislation, in particular the Work at Height Regulations and the Management of Health and Safety at Work Regulations.

A method statement for working at height should include:

▶ methods used to prevent falls;
▶ detailed breakdown of the job and any equipment to be used; and
▶ the name of the immediate supervisor.

203

Protection against falls

When working at height the following hierarchy of measures should be considered to prevent falls:

- ▶ avoid working at height, if possible;
- ▶ the provision of a properly constructed working platform, complete with toe boards and guard rails;
- ▶ if this is not practicable or where the work is of short duration, suspension equipment should be used; and only when this is impracticable,
- ▶ collective fall arrest equipment (air bags or safety nets) may be used;
- ▶ where this is not practicable, individual fall restrainers (safety harnesses) should be used; and
- ▶ only when none of the above measures are practicable, should ladders or step ladders be considered.

Protection from falling objects could be provided by:

- ▶ the use of covered walkways or suitable netting to catch falling debris
- ▶ waste material brought to ground level by the use of chutes or hoists
- ▶ minimal quantities of building materials stored on working platforms
- ▶ hard hats worn by all employees, contractors and visitors.

The effect of deteriorating materials and weather

- ▶ ageing materials become brittle, weaker and less stable
- ▶ high winds can affect the stability of scaffolding and ladders
- ▶ high winds can displace tools and materials and cause injuries to persons working at or below the workings
- ▶ lightning is another weather hazard to those working at height
- ▶ increased risk of skin cancer due to excessive exposure to sunlight
- ▶ cold weather can also present hazards when working at height. Guard rails become colder and the reactions of workers much slower.

Emergency rescue plans

▶ simple and straightforward rescue method;
▶ proprietary schemes available for complex rescues;
▶ any rescue system selected needs to be proportionate to the risk;
▶ there should not be undue reliance on the emergency services. However, arrangements should be in place to inform the emergency services of any serious incident;
▶ instruction and training are essential for the teams, preferably by simulated rescue exercises;
▶ trained rescue teams available at all times with a designated leader; and
▶ awareness of suspension trauma.

Requirements for head protection

The Construction (Head Protection) Regulations virtually mandated employers to supply head protection (hard hats) to employees whenever there is a risk of head injury from falling objects. Other requirements:

▶ the employer must ensure that hard hats are properly maintained and replaced when they are damaged;
▶ self-employed workers must supply and maintain their own head protection;
▶ visitors to construction sites should always be supplied with head protection; and
▶ mandatory head protection signs displayed around the site.

Head protection is now included in the PPE Regulations.

Inspection requirements for work equipment

The reporting requirements for work equipment inspections are:

▶ name and address of the person for whom the inspection was undertaken;
▶ date and time of the inspection;
▶ location of the work equipment inspected;
▶ description of the work equipment inspected;
▶ date and time of the inspection;

205

- details of any matter identified that could give rise to a risk to the health or safety of any person;
- details of any action taken as a result of any matter identified in the previous paragraph;
- details of any further action considered necessary; and
- name and position of the person making the report.

Safe working practices for access equipment and roof work

Fixed scaffold

Fixed scaffolds are usually independently tied and important considerations are:

- scaffolding must only be erected and dismantled by competent people – any changes to the scaffold must be made by a competent person;
- in busy town centres, the erection and dismantling of the scaffold should take place during quiet times;
- the scaffold must be designed to carry all the required loads and use only sound fittings and materials;
- adequate toe boards, guard rails and intermediate rails must be fitted to prevent people or materials from falling;
- the use of appropriate bay lengths and widths, bracing and tie-in arrangements together with edge and stop-end protection;
- the scaffold must rest on a stable surface; uprights should have base plates and timber sole plates if necessary;
- the scaffold must have safe access and egress;
- work platforms should be fully boarded with no tipping or tripping hazards;
- the scaffold must be maintained and cleaned and have proper arrangements for waste disposal;
- the scaffold should be sited away from or protected from traffic routes so that it is not damaged by vehicles;
- lower level uprights should be prominently marked in red and white stripes;

▶ if the scaffold must be erected near to vehicle roadways then barriers or fenders should be fitted around its base;

▶ the scaffold should be properly braced and secured to the building or structure;

▶ overloading of the scaffold must be avoided;

▶ the public must be protected at all stages of the work; and

▶ regular inspections of the scaffold must be made and recorded.

There are two basic types of external scaffold:

▶ independent tied

▶ putlog.

Components of a scaffold:

▶ Standard

▶ Ledger

▶ Guard rail

▶ Toe boards

▶ Bracing

▶ Transom

▶ Base plate

▶ Sole board

▶ Ties

▶ Working platform.

Factors affecting the stability of an independently tied scaffold include:

▶ poor, incompetent erection and/or a lack of regular inspections;

▶ the strength of the supporting ground or foundation;

▶ the proximity to the scaffold of any excavation work;

▶ the effect of excessive surface water in weakening the scaffold foundation;

▶ unauthorised alteration by incompetent persons (perhaps by vandals);

▶ the use of incorrect or damaged fittings during the erection or extension of the scaffold;

▶ the overloading of the scaffold;

▶ adverse weather conditions;

▶ the sheeting in of the scaffold without the use of extra ties; and
▶ vehicular impact.

Other points on scaffolds include:

▶ Brick guards and debris netting should be used to prevent materials and equipment falling to the ground. Brick guards are the best means of preventing materials from falling over the toe board on a scaffold.
▶ The maximum permitted gap between guard rails on a working platform is 470 mm.
▶ For putlog scaffolds, the putlog ends should be fully inserted and securely bedded into well-constructed brickwork.
▶ A putlog scaffold should not exceed 45 m in height.
▶ All scaffolds should have an assembly, use and dismantling plan.
▶ Only an authorised and competent person should erect, dismantle or alter a scaffold.
▶ Before accepting for use a scaffold erected by a specialist scaffolding sub-contractor, the site supervisor must check the scaffold and obtain a handing-over certificate.
▶ A scaffold must be inspected by a competent person after it is erected and at intervals not exceeding 7 days.
▶ The site supervisor, at the beginning of each week, should inspect the scaffold and ensure that any defect is rectified.
▶ A report of the inspection and action taken should be entered in the site inspection register.
▶ A similar inspection should also be carried out after high winds or other adverse weather conditions.
▶ Any scaffold that is being erected, altered or dismantled, or otherwise not suitable for use by workers, must have a warning notice erected stating that it must not be used.
▶ All scaffolds must be checked at the end of each working day to ensure that access to the scaffold by children has been prevented.

The requirements of scaffold erectors are that they should:

▶ be competent and experienced scaffolders;
▶ have a good reputation and accident record for scaffolding work;

▶ be supervised throughout the erection of the scaffold;

▶ be capable of inspecting scaffolding materials and equipment before they are used;

▶ be equipped with personal lifelines which must be attached at all times while working at height; and

▶ have attended a specialist scaffolding course and be able to provide documentary evidence that they have successfully completed the course.

Design of loading platforms involves the following features:

▶ the platform is built by laying scaffold boards across the transoms and, normally, at least three support points are required for each board;

▶ there must be a slight overhang (50 to 150 mm) across a support at each end of the board;

▶ boards are normally butt jointed;

▶ toe boards and guard rails, at least 950 mm above the working platform, must be fitted inside the standards along the edge of the working platform;

▶ ladders used to access the loading platform must extend at least 1 m above the level of the platform;

▶ all stacking of materials and equipment, such as bricks, must be as close to standards as possible to enable the added weight to be more evenly distributed throughout the scaffold;

▶ the boards need some additional fixing to the structure so that they are protected from lifting during high winds; and

▶ the width of the platform varies with the type of work being undertaken, from a minimum of three boards to a maximum of seven (when trestles are to be used on the platform).

Relevant points for the design of all working platforms, whether loading or not, are:

▶ they must be wide enough (at least 600 mm) to allow people to pass each other and for any required equipment or materials to be carried and stored;

▶ they must be free of openings and trip hazards;

▶ if members of the public are able to pass beneath, a double-boarded platform with a sheet of polythene placed between the

209

boards may be necessary to prevent debris or tools falling on them;

▶ the principal hazards associated with working platforms are overloading, the fall of materials, tools and equipment, and slips, trips and falls of people working on the platform; and

▶ good housekeeping and supervision are essential on all working platforms.

If materials are stored on a working platform, then:

▶ the platform must be strong enough;

▶ the platform must not become unsafe for other workers; and

▶ the materials must be secure and not be moved by strong winds.

▶ A working platform must be inspected before its first use and after every 7 days.

Scaffold hoists

All hoists must be:

▶ well maintained, inspected regularly and in good working order;

▶ positioned on firm and stable ground;

▶ erected by trained and experienced people who will follow the manufacturer's instructions and ensure that the hoist is properly secured to the supporting structure;

▶ operated by a driver who is properly trained and competent;

▶ properly protected at either end; and

▶ marked with safe working load and control functions.

The main hazards associated with hoist operations are:

▶ failure of equipment due to lack of maintenance;

▶ incorrect positioning of lifting appliance;

▶ insecure attachment of load and falls from the platform;

▶ unstable slinging of loads (hook not over centre of gravity);

▶ contact with overhead electricity cables;

▶ lack of operator training;

▶ no banksman used when driver's view is obscured; and

▶ incorrect signals used.

The following additional precautions may be needed:

▶ the controls set so that the hoist is operated from one position only (normally ground level)

▶ all the landing levels must be clearly visible by the operator from the operating position

▶ the hoistway must be enclosed at working platforms or window openings

▶ gates must be provided at all landings and at ground level

▶ the gates at landings are kept closed except during loading and unloading

▶ there must be no significant gap between the hoist platform and the landing

▶ loads, such as wheelbarrows, should be securely chocked and not overfilled

▶ loose loads, such as bricks, should be carried in proper containers or a hoist with an enclosed platform should be used

▶ the platform must not be overloaded and should be clearly marked with its safe working load

▶ loads must be evenly distributed on the hoist platform

▶ the hoist must be thoroughly examined and tested after erection, substantial alteration or repair and at 6-month intervals

▶ regular inspections should be carried out at least on a weekly basis and the results recorded.

Pre-fabricated mobile scaffold towers

The following points must be considered when mobile scaffold towers are to be used:

▶ towers should only be used on level, firm and stable ground;

▶ the selection, erection, dismantling and inspection of mobile scaffold towers must be undertaken by competent and trained persons with maximum height to base ratios not being exceeded;

▶ no person must be permitted to erect, alter or dismantle any mobile tower scaffold unless authorised by the site supervisor;

▶ diagonal bracing and stabilisers should always be used;

▶ access ladders must be fitted to (not leaning on) the narrowest

side of the tower or inside the tower and persons should not climb up the frame of the tower;

▶ all wheels must be locked whilst work is in progress and all persons must vacate the tower before it is moved;

▶ all operatives who are required to use mobile tower scaffolds will be instructed in their safe use and movement;

▶ outriggers or stabilisers must be extended where applicable;

▶ tower scaffolds must not be used or moved on sloping, uneven or obstructed surfaces;

▶ tower scaffolds must only be used in the vertical position;

▶ the tower working platform must be fully boarded, fitted with guard rails and toe boards and not overloaded;

▶ towers must be tied to a rigid structure if exposed to windy weather or to be used for work such as jet blasting;

▶ persons working from a tower must not over-reach or use ladders from the work platform;

▶ safe distances must be maintained between the tower and overhead power lines both during working operations and when the tower is moved; and

▶ the tower should be inspected on a regular basis and a report made.

The main reasons why a mobile tower scaffold may become unstable are:

▶ the brakes are defective;

▶ it was not erected by a competent person or unauthorised adjustments were made;

▶ it was erected on ground that was neither firm nor level;

▶ the height of the tower exceeded the recommended base to height ratio;

▶ the failure to fit and/or use outriggers when they were required;

▶ the safe working load was exceeded;

▶ the movement of the tower with either persons and/or materials on the working platform or under windy conditions; and

▶ collision with vehicles or other plant.

Mobile Elevating Work Platforms (MEWPs)

The following factors must be considered when using MEWPs:

▶ the MEWP must only be operated by trained and competent persons;
▶ safety harnesses should be worn;
▶ it should be inspected before use;
▶ the working platform should not be overloaded;
▶ it must never be moved in the elevated position;
▶ it must be operated on level and stable ground with consideration being given to the stability and loading of floors;
▶ the tyres must be properly inflated and the wheels normally immobilised when the platform is in use;
▶ outriggers, if fitted, should be fully extended and locked in position;
▶ warning signs should be displayed and barriers erected to avoid collisions;
▶ due care must be exercised with overhead power supplies and obstructions and adverse weather conditions;
▶ it should be maintained regularly and procedures should be in place in the event of machine failure; and
▶ drivers of MEWPs must be instructed in emergency procedures, particularly to cover instances of power failure.

Ladders

The following factors should be considered when using ladders:

▶ aluminium ladders are light but should not be used in high winds or near live electricity;
▶ timber ladders need regular inspection for damage and should not be painted as this could hide cracks;
▶ ensure that the use of a ladder is the safest means of access for the work to be done and the height to be climbed;
▶ the ladder needs to be stable in use with a safe inclination (1 in 4);
▶ the best way to prevent a ladder from slipping is to tie it at the top;

213

▶ if possible, the foot of the ladder should also be tied to a rigid support;
▶ weather conditions must be suitable;
▶ the proximity of live electricity should be checked;
▶ the ladder should be fitted with non-slip feet;
▶ there should be at least.1 metre of ladder above the stepping off point;
▶ over-reaching must be eliminated;
▶ workers who are to use ladders must be trained in the correct method of use and selection; and
▶ ladders should be inspected (particularly for damaged or missing rungs) and maintained on a regular basis, and they should only be repaired by competent persons.

Certain work should not be attempted using ladders. This includes work where:

▶ two hands are required;
▶ the work is at an excessive height;
▶ the ladder cannot be secured or made stable;
▶ the work is of long duration;
▶ the work area is very large;
▶ the equipment or materials to be used are heavy or bulky;
▶ the weather conditions are adverse; and
▶ there is no protection from vehicles.

Causes of accidents with any access equipment

▶ Design faults
▶ Over-reaching
▶ Falls and slips
▶ Collision with vehicles
▶ Unsound base support
▶ Untrained personnel
▶ Climbing with loads
▶ Poor maintenance and/or inspection
▶ Instability
▶ Adverse wind effects.

Other access techniques

Boatswain's chair

▶ used for light work of a short-term nature
▶ the worker should be protected by a harness and lanyard.

Cradles

Suspended access cradles or platforms are used for maintenance operations on bridges or similar inaccessible structures. The reasons for accidents involving cradles include:

▶ poor equipment selection procedures – cradle is not fit for purpose
▶ unsafe access to the cradle
▶ overloading of the cradle or non-uniform loading across the platform
▶ holes or cracks in the platform floor allowing material to fall below
▶ inadequate guard rails and toe boards
▶ the structure supporting the cradle is not capable of carrying the additional load
▶ insecure counterweights and/or braking system
▶ the failure of essential components of the cradle structure
▶ the failure of the winching and climbing devices
▶ poor erection, maintenance and dismantling by incompetent and untrained personnel
▶ inadequate emergency procedures.

Other points:

▶ The safest access to a cradle is at ground level.
▶ If access at height is unavoidable, then the cradle must be secured to prevent it swinging away from the structure.
▶ Suitable anchorage points must be available for safety harnesses and a secondary rope fitted.
▶ Regular visual inspections must be made before each time it is used.
▶ A weekly detailed inspection must be made by a competent person.

Rope access

▶ Abseiling or rope access techniques are used for inspection work when a properly constructed working platform is not practicable.

▶ Rope access must be erected by a competent person and the worker must be properly trained in the use of the technique.

▶ The main rope and safety access rope must always be attached to separate anchorage points.

▶ Any tools or equipment used must be secured to the worker to prevent them falling.

▶ The area beneath the work should be fenced off.

Fall arrest equipment

The three most common types are:

Safety harnesses

▶ The last resort for workers when conventional protection, using guard rails, is no longer practicable.

▶ Safety harness and lanyard is a fall arrest system.

▶ Only the length of fall is reduced and the worker may still be injured due to the shock load applied to him/her when the fall is arrested.

▶ The lanyard should always be attached above the worker, whenever possible.

▶ The worker must be attached to a secure anchorage point before he/she moves into an unsafe position.

▶ Only trained and competent workers should attach lanyards to anchorage points and work in safety harnesses.

▶ Those who wear safety harnesses must be able to undertake safety checks and adjust the harness before it is used.

Safety nets

▶ Safety nets arrest falls of people, tools and materials from height.

▶ Competent installation is essential.

216

▶ The correct tensioning of the net is important.
▶ Specialist companies normally fit nets.
▶ Nets are not suitable for use in low-level construction as there is insufficient clearance below the net.
▶ Nets should be positioned so that workers will not fall more than 2 m or hit the ground or other obstructions.

Air bags

▶ Air bags are used when it is not practical to use safety nets.
▶ The bags must be of sufficient strength and air pressure.
▶ The inflation pump must run while work is in progress.

If it is not possible to install edge protection or a soft landing system, then a harness and lanyard should be worn and anchored to a fixed point.

Roof work

The main hazards are:

▶ fragile roofing materials – more brittle with age and exposure to sunlight
▶ exposed edges
▶ unsafe access equipment
▶ fragile roof lights and voids
▶ falls from girders, ridges or purlins
▶ overhead services and obstructions
▶ the use of equipment such as gas cylinders and bitumen boilers
▶ manual handling hazards.

The controls include:

▶ a risk assessment and a method statement must be completed before any roof work takes place;
▶ suitable means of access such as scaffolding, ladders and crawling boards;
▶ suitable barriers, guard rails or covers where people work near to fragile materials and roof lights;

217

- ▶ edge protection on a flat roof is best provided by double guard rails and toe board;
- ▶ suitable warning signs indicating that a roof is fragile, at ground level;
- ▶ access to the area immediately below the work should be restricted using suitable barriers, netting, safety signs and safety helmets;
- ▶ the safe transportation of materials to and from the working area, involving lifts, hoists, manual handling and/or using chutes and covered waste skips; and
- ▶ good housekeeping procedures.

A risk assessment for roof work should include:

- ▶ an assessment of the structural integrity of the roof;
- ▶ the methods to be used to repair the roof; and
- ▶ a COSHH assessment of any hazardous substances to be used.

Protection of others

Members of the public must be protected during construction work. The protection methods include:

- ▶ the painting of scaffold standards with red and white stripes;
- ▶ masking couplers;
- ▶ adequate warning signs;
- ▶ good illumination at night;
- ▶ the provision of barriers and other means of safe passage;
- ▶ sheeting, netting and fans;
- ▶ the provision of head protection for visitors; and
- ▶ the boarding of ladders when the site is unoccupied.

When construction work is being undertaken on occupied premises, the following issues may arise:

- ▶ the location and isolation of the construction working areas;
- ▶ supervision of the work;
- ▶ procedures for the emergency evacuation of the premises;
- ▶ the use of welfare and canteen facilities;
- ▶ the protection of the occupier's employees;

▶ procedures for handing back areas of the store when the work has been completed; and

▶ the safe removal of all waste materials.

Working over water

Hazards

▶ falling from a height

▶ scaffold stability

▶ falling material

▶ pedestrian safety

▶ possible collision with the scaffolding by boats

▶ leptospirosis and tetanus.

Possible controls

Include those for working at height and:

▶ the provision of rescue equipment such as life belts and grab lines;

▶ the availability of a rescue boat under the control of a competent person, should someone fall into the water;

▶ harnesses must be worn by those working on the scaffold in addition to the use of appropriate personal protective equipment;

▶ there should be a current register of workers and nobody should work alone on the scaffold;

▶ for pedestrians, using perhaps a towpath, a protected access or safe diversion route must be provided;

▶ signs, lighting and protective barriers will also be required to minimise the risk of boats colliding with the scaffold; and

▶ disinfection, good standards of personal hygiene, the use of appropriate personal protective equipment, adequate washing and first-aid facilities.

At the planning stage the following points need to be considered:

▶ the provision of barriers or fences with toe boards;

▶ the provision of life jackets, buoyancy aids and grab lines;

219

▶ the provision of rescue equipment, lifebelts, harnesses, safety lines and, in some cases, a rescue boat;

▶ if transport by water is required, arrangements must be made for safe landing stages and life jackets;

▶ any boats that are used should be of suitable construction; under the control of a competent person; not overloaded and properly maintained;

▶ the provision of rescue teams and procedures;

▶ training for boat operatives, rescue teams and supervisors; and

▶ precautions to protect the safety of the public, especially children.

The personal protective equipment for working over water is a life jacket, harness and lanyard.

2.11

Excavation work and confined spaces – hazards and risk control

Learning outcomes	
Explain the hazards and risk assessment of excavation work	☐
Explain the control measures for excavation work	☐
Explain the hazards and risks associated with confined space working	☐
Outline the control measures for confined space working	☐

Excavations: work hazards and assessment

Excavation hazards

▶ premature collapse of the sides (possibly due to soil composition);
▶ materials falling on workers;
▶ excavation machinery and equipment (e.g. pneumatic drills);
▶ people and/or vehicles falling into the excavation;
▶ workers struck by plant;
▶ manual handling;
▶ noise and vibration;
▶ hazardous substances, particularly near the site of current or former industrial processes;
▶ influx of ground or surface water and entrapment in silt or mud;
▶ proximity of stored materials, waste materials or plant – surcharging;
▶ proximity of adjacent buildings or structures and their stability;
▶ contact with underground services;
▶ contact with overhead power lines;
▶ access and egress to the excavation; and
▶ dust, fumes, lack of oxygen and other health hazards (such as Weil's disease).

Site survey should include:

▶ previous use of site;
▶ location of existing buildings;
▶ location of new structures;
▶ results of soil investigations;
▶ ground contamination;
▶ level of water table and type of soil;
▶ storage and disposal of excavated material;
▶ amount of working space and storage required;
▶ the most suitable method of temporary support of the excavation walls; and
▶ adequate emergency arrangements.

Cofferdams and caissons

Some of the more common uses for **cofferdams** include:

▶ bridge construction;
▶ hydroelectric dam construction;
▶ ship repair – cofferdams are used to generate a 'dry dock' condition; and
▶ the recovery of sunken ships in shallow waters.

Some of the advantages of cofferdams are that:

▶ they allow excavation and construction of structures in difficult environments;
▶ they provide a safe workplace;
▶ the steel sheet piles are easily installed and removed; and
▶ many of the materials can be reused.

Main types of **caisson** are:

▶ box caissons
▶ open caissons
▶ compressed-air or pneumatic caissons.

Some of the more common uses for **caissons** include foundations for buildings and bridges.

Hazards of cofferdams and caissons:

▶ flooding

- collision with shipping
- unstable soil
- weather and tides
- the hoisting of people and materials and construction work equipment.

Additional caisson hazards are:

- oxygen depletion
- toxic fumes
- fire.

There is also a potential health problem of working in the compressed air atmosphere of a caisson.

Risk assessments for excavations

Should consider the following factors:

- any alternatives to excavation?
- nature of the ground, soil structure and groundwater regime;
- location of the excavation (e.g. readily accessible public place, contaminated ground or heavily serviced urban area);
- depth of excavation;
- presence of hazardous atmospheres and contaminated soil;
- contact with existing underground or overhead services;
- nature of the work required to be undertaken within the excavation;
- required method of support of the excavation;
- removal from site of surplus excavated material;
- stockpiling of excavated material which is to be reused;
- fencing required to prevent plant and materials falling into excavations;
- provision of secure storage space for materials and working space for workers and plant;
- type of excavation and compaction equipment to be used;
- precautions necessary to prevent unauthorised access to the excavation;
- emergency arrangements, including first-aiders; and

▶ arrangements required for traffic management on adjacent public highways.

Control measures for excavation work

These include:

▶ supervision by a competent person;
▶ a competent person must inspect an excavation each time that a shift starts work;
▶ all excavation workers must be competent;
▶ safe access by ladders and crossing points for pedestrians and vehicles;
▶ hard hats, safety footwear and other relevant personal protective equipment must be worn;
▶ an excavation must be supported if there is a risk of the sides collapsing;
▶ erection of barriers around the top of the excavation;
▶ not storing spoil material near the top of the excavation;
▶ vehicles kept away from the top of the excavation using warning signs and barriers;
▶ if a vehicle is tipping materials into the excavation, stop blocks should be placed behind its wheels;
▶ control of any influx of water;
▶ care with use of freezing equipment;
▶ control of hazardous substances (e.g. exhaust fumes) or health hazards present;
▶ workings well lit at night;
▶ identify position of buried services; and
▶ care during the filling in.

Methods of support include:

▶ battered sides
▶ open and close sheeting
▶ hydraulic waling frames
▶ manhole shores
▶ trench boxes
▶ drag boxes.

The location of underground services requirements:

▶ accurate service plans;

▶ excavation workers trained in safe digging practices and emergency procedures;

▶ locators used to trace any services, where possible, or hand dig a series of trial holes with insulated tools;

▶ assume that all pipes or cables are 'live' unless it is known otherwise;

▶ report any suspected damage to services to the service providers and health and safety enforcement agency; and

▶ update service plans when the new services have been laid.

Inspections:

▶ every 7 days with a report

▶ every day:

▷ after any event likely to affect the stability of the excavation

▷ before the start of each shift

▷ after an accidental fall of material.

The controls for **cofferdams** are:

▶ must be designed by a qualified engineer and constructed and maintained in accordance with the design;

▶ if the flooding of the cofferdam by high waters is possible, then means must be provided for controlled flooding of the work area;

▶ emergency evacuation procedures of workers must be posted and communicated to the workforce;

▶ there should not be fewer than two means of egress; and

▶ if they are located close to navigable shipping, then warning devices that are visible to passing ships must be provided.

The controls for **caissons** are:

▶ they must be designed by a qualified engineer and constructed and maintained in accordance with the design;

▶ each employee who is required to enter a caisson excavation must have a lanyard attached to a body harness;

▶ all employees shall be removed from the caisson when material is being hoisted from the caisson;

226

▶ a supervisor must be stationed at the caisson whenever workers are present;

▶ a secondary mechanical means that is capable of removing an employee from the caisson shall be readily available should the crane performing the caisson work break down;

▶ the air quality of a caisson must be tested at regular intervals;

▶ emergency evacuation procedures of workers must be posted and communicated to the workforce;

▶ if they are located close to navigable shipping, then warning devices that are visible to passing ships must be provided;

▶ caissons shall be substantially braced before loading with concrete or other weight; and

▶ when two or more caissons are sunk together, employees shall be removed from adjacent caissons during concreting operations.

Confined space working hazards and risks

The Confined Spaces Regulations require that:

1 No person at work shall enter a confined space for any purpose unless it is not reasonably practicable to achieve that purpose without such entry.

2 Other than in an emergency, no person shall enter, carry out work or leave a confined space otherwise than in accordance with a safe system of work relevant to the specified risks.

Definition of a confined space:

A confined space can be defined as any space of an enclosed nature which has limited means of access and egress, restricted natural ventilation and is not intended for continual occupancy by persons.

Examples of a confined space:

Manholes, sewers, tunnels, excavations, storage tanks, holds of ships, pits, trenches, ducts, some unventilated areas or rooms within buildings, particularly below ground level, boilers, combustion chambers in furnaces, chambers, vats, silos, pipes, flues and wells.

Hazards include:

▶ difficult access and egress;

- ▶ accumulation of vapours, gases or fumes;
- ▶ lack of ventilation;
- ▶ asphyxiation due to oxygen depletion;
- ▶ poisoning by toxic substance or fumes;
- ▶ explosions due to gases, vapours and dust;
- ▶ fire due to flammable liquids, vapours and oxygen enrichment;
- ▶ fall of materials leading to possible head injuries;
- ▶ free-flowing solid, such as grain in a silo;
- ▶ electrocution from unsuitable equipment;
- ▶ difficulties of rescuing injured personnel;
- ▶ drowning due to flooding;
- ▶ excessive heat leading to heat stress;
- ▶ claustrophobic effects due to restricted space; and
- ▶ diseases from animal wastes, infected materials or micro-organisms.

Risk assessment should include:

- ▶ the physical dimensions and layout of the confined space;
- ▶ any oxygen deficiency or enrichment;
- ▶ the use of cleaning chemicals and any interaction with other substances;
- ▶ any residues that have been left in the confined space (e.g. sludge) and details of gases or vapours which may be generated if these substances are disturbed;
- ▶ any contamination which may arise from adjacent plant, processes, services, pipes or surrounding land, soil or strata;
- ▶ any sources of ignition for flammable dusts, gases or vapours;
- ▶ the need to isolate the confined space from outside services or from substances held inside the space; and
- ▶ emergency rescue arrangements, including trained people and equipment.

Control measures for confined space working

The main elements of a safe system of work (possibly a 'permit to work') are:

- ▶ extent of the supervision of the work;

- competence, training and instruction of the workforce;
- stipulation of a minimum gang size for large confined spaces;
- details of any information required by the workforce;
- methods of communication between people inside, from inside to outside and to summon help in an emergency;
- testing and monitoring of the atmosphere inside the confined space for hazardous gas, fume, vapour, dust and the concentration of oxygen;
- gas-purging of toxic or flammable substances with air or inert gas;
- provision of good ventilation, possibly by mechanical means;
- if appropriate, the cleaning of the space before the work commences;
- removal of residues using appropriate equipment;
- effective and complete isolation of electrical and mechanical equipment by the use of a lock-off and a tag system with key security;
- provision of personal and respiratory protective equipment if it is not possible to enter the confined space otherwise;
- access and egress to the confined space in an emergency;
- details of fire prevention equipment and procedures;
- lighting arrangements, including emergency lighting;
- emergency and rescue arrangements and procedures (including first-aid);
- maximum length of any one working period;
- details of any required health surveillance of the workforce; and
- monitoring and audit arrangements.

Monitoring arrangements

- before any work commences, the supervisor must arrange for any necessary environmental surveys and sampling needed to protect the health of the workforce;
- competent specialists required to test and monitor the atmosphere and interpret the results accurately;
- continuous gas monitoring should be used over the complete work period;

- regular checks on all breathing and reviving apparatus, rescue equipment and emergency alarms;
- regular checks on all first-aid provision and the accreditation of first-aiders; and
- periodic health checks given to the workforce by an occupational health specialist, and consideration should be given to the exclusion from confined spaces of individuals who have any of the following problems:
 - ▷ heart disorders, including high blood pressure
 - ▷ fainting, blackouts, fits or loss of balance
 - ▷ weak eyesight or hearing
 - ▷ asthma or chronic bronchitis
 - ▷ physical size, poor mobility or back pain; and/or
 - ▷ claustrophobia or psychological stress.

Emergency arrangements should incorporate the following:

- details of rescue teams and their individual responsibilities;
- contact names and telephone numbers of the local emergency services;
- details of the communication arrangements from inside the confined space;
- provision and maintenance of the rescue and resuscitation equipment;
- raising of the alarm, alerting the rescue team and maintaining close supervision of the workforce inside the confined space;
- notification of and consultation with the emergency services;
- the protection and safeguarding of the rescue team;
- fire safety precautions and procedures;
- provision and maintenance of fire-fighting equipment;
- control and possibly the protection of adjacent buildings, plant and equipment;
- the arrangements for dealing with emergencies outside the normal working day;
- first-aid arrangements, including any special training required of first-aiders;
- training of rescuers and simulations of emergencies, such as fire drills; and

▶ size of access openings to permit rescue with the normal respiratory protection and rescue equipment for confined spaces.

General points

▶ Work in a confined space requires a risk assessment, a method statement and a permit to work.

▶ The first consideration when planning a confined spaces permit to work is whether there is an alternative method of working.

▶ Before any work starts in a confined space, a rescue plan and team must be in place.

▶ The top person at a confined space begins the rescue if it is needed.

▶ A confined space must be vacated if the permit to work expires.

Demolition and deconstruction – hazards and risk control

Learning outcomes	
Identify the main hazards of demolition and deconstruction work	☐
Outline the control measures for demolition and deconstruction work	☐
Identify the purpose and scope of a pre-demolition/ refurbishment survey	☐
Outline the main control measures that a demolition/ refurbishment method statement should include	☐

🔑 **Key revision points**

The meaning of deconstruction, piecemeal demolition and deliberate controlled collapse	☐
The hazards and controls required for demolition and deconstruction work	☐
The contents of a demolition method statement	☐
The details of a pre-demolition survey	☐

Demolition and deconstruction hazards and risks

Deconstruction preserves various components of the structure, such as wood and bricks, for reuse at another building or structure.

Demolition reduces the structure to rubble, which is either treated as waste or, in some cases, used as hard core for other construction projects.

There are two forms of demolition:

▶ piecemeal
▶ deliberate controlled collapse.

Principal hazards of demolition and deconstruction work

▶ falls from height or on the same level;
▶ falling debris;
▶ premature collapse of the structure being demolished;
▶ dust and fumes;
▶ the silting up of drainage systems by dust;
▶ the problems arising from spilt fuel oils;
▶ manual handling;
▶ presence of asbestos and other hazardous substances;

234

▶ noise and vibration from heavy plant and equipment;
▶ electric shock;
▶ fires and explosions from the use of flammable and explosive substances;
▶ smoke from burning waste timber;
▶ pneumatic drills and power tools;
▶ the existence of services, such as electricity, gas and water;
▶ collision with heavy plant; and
▶ plant and vehicles overturning.

Control measures

Management of the demolition (and deconstruction) and general controls include all or some of the following:

▶ avoidance of premature collapse;
▶ disconnection of services to the site;
▶ separation between the public and demolition activities using barriers and fences;
▶ excluding unauthorised persons from the site;
▶ use of an induced collapse technique, in place of people working at heights;
▶ use of power shears, in place of grinding or oxyacetylene cutting, where a risk of fire exists;
▶ provision of shatterproof/guarded windscreens on plant;
▶ provision of falling-object protective structures (FOPS) and rollover protective structures (ROPS) on vehicles and plant;
▶ installation of safe working platforms on scaffolds and elevated work platforms;
▶ installation of edge protection to open edges of landings, stairways and fixed platforms;
▶ installation of static lines and anchor points for fall arrest systems;
▶ use of chutes for dropping debris;
▶ use of cranes for lowering loads;
▶ limiting the amount of time a person is exposed to a particular hazard (e.g. vibration); and

▶ training and instruction for all workforce (including an explanation of the method statement).

A risk assessment informs the contents of the method statement.

Demolition method statement should include:

▶ method of demolition to be used, including prevention of premature collapse;
▶ safe removal of debris from upper levels;
▶ site access and security;
▶ location of any underground or overhead services;
▶ arrangements for protection from falling materials;
▶ equipment, including access equipment, required;
▶ any hazardous substances to be used;
▶ arrangements for the protection of the public and the workforce against noise and release of hazardous substances;
▶ details of the isolation methods for any services which may have been supplied to the site and any temporary services required on the site;
▶ details of personal protective equipment to be worn on site;
▶ first-aid, emergency and accident arrangements;
▶ training and welfare arrangements;
▶ arrangements for waste disposal;
▶ names of site supervisors; and
▶ required level of competence of site workers.

Summary of demolition controls:

▶ prior to work, a full site investigation;
▶ risk assessments (two required – project designer and contractor);
▶ a method statement;
▶ waste disposal arrangements;
▶ piecemeal and deliberate controlled collapse; and
▶ training and information to all workers.

Personal protective equipment includes:

▶ Safety helmets
▶ Fall arrest harnesses and lanyards
▶ Safety boots

236

▶ Gloves
▶ Goggles
▶ Respirators
▶ Hearing protectors.

Purposes and scope of pre-demolition, deconstruction or refurbishment survey

The property owner (the client) has a duty to provide pre-demolition information to the Principal Designer and the Principal Contractor.

Before any work is started, a full site investigation must be made by a competent person (e.g. a specialist structural engineer) to determine the hazards and associated risks which may affect the demolition workers and members of the public.

Pre-demolition investigation and survey includes:

▶ construction details of the structures or buildings to be demolished and those of neighbouring structures or buildings;
▶ the identification of key structural elements, including pre- and post-tensioned components;
▶ a review of drawings, structural calculations, health and safety file related to the structure;
▶ previous use of the premises;
▶ load-carrying capacity of adjoining land, including the presence of underground culverts;
▶ need for possible temporary support structures for the building being demolished and adjoining buildings;
▶ prevention of falls of materials and people;
▶ location of any dangerous machinery;
▶ presence of asbestos, lead or other hazardous or radioactive substances;
▶ environmental issues, such as dust, water pollution and noise;
▶ public safety, including the provision of high fencing or hoardings;
▶ manual handling issues;
▶ location of any underground or overhead services (water, electricity, gas and sewage);

- ▶ location of any underground cellars, storage tanks, chimneys;
- ▶ means of access to the site;
- ▶ removal of waste;
- ▶ details of any traffic or pedestrian routes through the site;
- ▶ provision of welfare facilities;
- ▶ proximity of neighbours;
- ▶ location of any public thoroughfares adjacent to the structure or building; and
- ▶ name of the principal designer.

Specimen answers to NEBOSH-type examination questions

Introduction

This chapter is only relevant to those readers who are due to sit for NEBOSH examinations.

The accompanying textbook to this Revision Guide – *Introduction to Health and Safety in Construction* – gives detailed advice on studying for NEBOSH examinations and some sample solutions to practice examination questions based on NEBOSH examination questions. It also provides a sample practical assessment. To gain the full benefit from this section of the Revision Guide, those chapters of the textbook should also be read. But full success in the examinations is only likely to be achieved if a good examination technique is adopted.

Features of a good examination technique

A. Mnemonics

The use of mnemonics during your revision period can be very helpful in the examination. A mnemonic can be a sentence or a word that enables a person to remember a list of items. For example, the colours of the rainbow – red, orange, yellow, green, blue, indigo and violet – can be remembered by the sentence 'Richard Of York Gained Battles In Vain'. Similarly, the main factors in a manual handling assessment can be remembered by the word TILE – Task, Individual, Load and Environment.

The best mnemonics are those which students have devised for themselves. The reader is advised to invent his or her own set of mnemonics based on familiar words or sentences. Some examples of useful mnemonics are given below.

1. Management of health and safety

Elements in HSG65	Individual factors	Common topics in the management paper
Plan	**S**elf-interest	**S**afety culture
Do	**H**earing/memory loss	**T**raining
Check	**E**xperience	**R**isk assessment
Act	**A**ge	**I**nformation
	Training level	**M**aintenance
	Health	**S**upervision
		Safe system of work

SHEATH and **STRIMSS** are stand-alone words. (STRIMSS was devised by a student from the horticultural industry.) The reader should devise a suitable sentence to link the initial letters PDCA of HSG65.

2. Managing and controlling hazards in construction activities

Risk assessment stages (inc. COSHH and noise)	Machine hazards	Hierarchy of control for working at height
Assess the hazards	Entanglement	Avoid working at height if possible
Control the risks	Nips	Existing safe place of work used
Monitor the controls	Traps	Prevent falls by providing work equipment (e.g. guard rails)
Inform employees	Impact	Mitigate distance and consequences of a fall (e.g. airbags)
Record and review the assessment	Contact or cutting	Instruction, training and supervision
	Ejection	

ACMIR can be remembered by 'All Colours Must Include Red'.
AEPMI can be remembered by 'All Existing Parts Must Interchange'.
ENTICE is a stand-alone word.

It is very important to stress that a mnemonic is only an aid to memory, and not all the elements in a mnemonic may be relevant to a particular question. Where all or parts of a mnemonic are relevant to a question, you will need to expand on it in your answer.

For example, if the mnemonic, such as STRIMSS, reminds you that 'training and information' are relevant to the question, you should add a brief description of the training and information that you have in mind. Very few marks can be awarded for simply stating 'training and information'.

B. Use of personal experience

The NEBOSH course covers much factual material which many students find difficult to remember. It is easier to remember this material if it can be related to personal experience and their own workplace. For example, most workplaces have some hazardous substances present, even if only as cleaning materials. These can be used to illustrate an answer involving hazardous substances (provided that the reference is relevant to the question!). Always try

to use familiar examples of hazards and controls when answering examination questions involving those hazards and controls.

C. Simple rules when taking the examination

1 Always arrive at least 10 minutes before the start of the examination.
2 Quickly read the whole paper before attempting any question.
3 Start each question on a separate sheet in the answer book and use both sides of the paper. The current NEBOSH answer book has the relevant question number printed on each page of the book.
4 Make brief notes at the beginning of the question page in the answer book.
5 Answer the easiest question first – this may not necessarily be the first one.
6 Read the question carefully and highlight the action verbs (e.g. outline, describe, state, etc.). These action verbs are printed in **bold letters** in the question on the examination paper. It is common for candidates to answer the question that they would like to answer rather than the actual one on the examination paper.
7 Answer **all** parts of the question.
8 Attempt **all** questions – you can only get zero marks for an unanswered question.
9 Time is very limited during the examination and answers should be brief and to the point. Do not write all that you know about the topic being questioned or pad out your answer. You should spend no more than 25 minutes on the long answer question (covering about a page and a half) and 8 minutes on each short answer question (normally covering half a page). Most examination failures are caused by not allowing sufficient time to answer all the questions.
10 Finally, the answers should be easy for the examiner to read in terms of layout and standard of writing.

Specimen answers based on NEBOSH examination questions

Some specimen answers are given below to questions based on past NGC1 and NCC1 examination questions. Detailed advice and an example of NCC2 – the Health and Safety Practical Application – are given in the textbook that accompanies this Revision Guide (*Introduction to Health and Safety in Construction*).

It is recommended that, to gain full benefit from the specimen answers, the reader should attempt the questions initially without consulting the answers. The answers given are generally longer than would be expected in the examination because they have covered all possible options that are available in answering a particular question.

Finally, two long answer questions are provided for NGC1 because candidates often have more difficulty with this paper.

1. NGC1 – The management of health and safety

Question 1

A manufacturing company has experienced a number of accidents that were due to poor selection and implementation of risk controls.

a **Identify FOUR** of the general principles of prevention. **(4)**
b **Give** the meaning of the term 'hierarchy of control'. **(2)**
c **Outline** how a typical hierarchy of control should be applied. **(8)**
d **Outline** the factors to be considered when selecting an appropriate risk control method. **(6)**

Answer:

a Four general principles of prevention are:
 1 the avoidance of risks;
 2 the evaluation of unavoidable risks;
 3 controlling hazards at source; and
 4 replacing the dangerous by less or non-dangerous alternatives.

 (The other principles are:

243

5 adapting work to the individual;

6 adapting to technical progress;

7 developing a coherent prevention policy;

8 giving priority to collective over individual protective measures; and

9 giving appropriate instructions to employees.)

b A series of actions arranged or organised according to rank or importance designed to control risks, which are considered in order of importance, effectiveness or priority of measures. It is designed to control risks that normally begin with an extreme measure of control and end with personal protective equipment as a last resort.

c The first stage of the hierarchy is to eliminate the risks by either designing them out or changing the process. The next stage is the reduction of the risks by the substitution of hazardous substances with others which are less hazardous. If this were not possible, then isolation would have to be considered, using enclosures, barriers or worker segregation. The application of engineering controls such as guarding, the provision of local exhaust ventilation systems, or the use of reduced voltage systems or residual current devices would follow, as would management controls such as safe systems of work, training, job rotation and supervision, with the final control measure being the provision of personal protective equipment such as ear defenders or respiratory protective equipment.

d Factors to consider when selecting risk control measures include the application of the hierarchy of control. The more serious hazards may be eliminated completely, for example by avoiding working at height with the installation of a permanent working platform with stair access. In some cases, less hazardous substances or processes may be substituted, such as the use of water-based rather than oil-based paints. The control of risks by isolating them or segregating people from the hazard is an effective control measure which is used in many instances; for example separating vehicles and pedestrians in workplaces. The rest of the hierarchy of controls may be applied in a similar way to that shown in part (c).

The evaluation of the level of risk involved, together with its likely frequency, the severity of its outcome and its extent, is another factor. Other relevant factors that should be considered include:

- ▶ the number and type of persons exposed to the risk;
- ▶ the environment in which the control is to be used;
- ▶ the requirements of the law and the recommendations contained in relevant standards and guidance;
- ▶ the existing level of competence of the persons involved, with the possible need for the provision of additional training and supervision; and
- ▶ the initial purchase and subsequent maintenance costs of the chosen risk control method.

Question 2

Following a reportable accident to a contractor while on an employer's premises, an investigation found that the employer had failed to provide health and safety information to the contractor. The employer was not able to demonstrate any source of competent health and safety advice.

a **Outline** the duties that the employer owes to the contractor under the Health and Safety at Work etc. Act 1974. **(4)**
b **Outline** the health and safety information that should have been provided to the contractor before work commenced. **(8)**
c **Outline** the factors that the employer should consider when selecting an individual to fulfil the role of health and safety practitioner. **(8)**

Answer:

a The duties contained in Section 3 of the Health and Safety at Work (HSW) Act require an employer to ensure that persons working on their premises who are not direct employees, such as contractors, are not put at risk by the activities being carried out at the premises. Additionally, Section 4 places duties on the employer, as a controller of premises, to ensure, for example, that the premises and its plant and equipment are safe and without risks to health for persons using them as a place of work.

245

(**Author comment:** *It is possible to argue that the CDM Regulations are relevant here because all Health and Safety Regulations are made as a result of Section 15 of the HSW Act. However, in NEBOSH examinations, specific mention would be made of any Regulation that the candidates were expected to include in their answers. Hence this question was only concerned with the specific duties included in the Act.*)

b The health and safety information that should be provided to the contractor before work commenced would include details of any hazards in the areas where they were to work, and details of any restricted areas and those where personal protective equipment such as hearing protection would have to be worn. Information would also be required on the emergency arrangements, such as for fire and first-aid, and accident (and hazard) reporting procedures. Any general site rules, for example for signing in and out of the site, should also be supplied, and the name and location of the designated contact person. Finally, information should be provided on any training procedures for any processes or equipment to be used by the contractor.

c The factors to be considered when selecting an individual to fill the role of health and safety practitioner include:

▶ previous training, qualifications and membership of professional bodies;

▶ knowledge and understanding of the role and relevant previous experience;

▶ understanding of the principles of risk assessment and relevant control methods;

▶ familiarity with relevant current legislation, standards and guidance;

▶ personal qualities, including ability to promote and communicate good health and safety practices;

▶ ability to identify problems and assess the need for action – a proactive approach;

▶ ability to develop and implement appropriate health and safety management strategies and evaluate their effectiveness using appropriate benchmarking data; and

▶ awareness of personal limitations and willingness to undertake training to rectify the deficiencies.

Question 3

Outline the general duties, under Section 6 of the Health and Safety at Work etc. Act 1974, of designers, importers, manufacturers and suppliers of articles and substances for use at work to ensure that they are safe and without risk to health. **(8)**

Answer:

Section 6 of the Health and Safety at Work etc. Act 1974 places an obligation on persons who design, manufacture, import or supply any article or substance for use at work to ensure, so far as is reasonably practicable, that it is safe and without risk to health. Articles must be safe when they are set, cleaned, used and maintained. Substances must be without risk to health when they are used, handled, stored or transported. Testing and examination must be carried out to ensure the required level of safety, and additionally employers should be provided with information on the safe use, dismantling and disposal of the articles and substances and given revised information should a subsequent serious risk become known. Importers have a duty to ensure articles or substances comply with the requirements of UK legislation.

Question 4

a **Identify FIVE** common law duties that an employer must fulfil. **(4)**

b **Outline** practical ways in which an employer might fulfil the five duties identified in (a). **(4)**

Answer:

a Five common law duties that an employer must undertake are:
▶ provide a safe place of work, including access and egress
▶ provide safe plant and equipment
▶ provide a safe system of work
▶ provide safe and competent fellow employees

247

▶ provide adequate levels of supervision, information, instruction and training.

b A safe place of work includes items such as the structure of the building and the condition of the flooring, both of which should be sound and free of defects. There should be a safe means of access and egress to the place of work, with gangways kept clear and wide enough for traffic if it is used. Environmental factors, such as working space, lighting and heating, are also important.

Safe plant and equipment will require adequate guarding or other safety devices and regular recorded maintenance.

Safe systems of work, such as permits to work and method statements, need to be available and communicated to the workforce.

Safe and competent fellow employees can be attained through pre-selection, induction and training programmes and good supervision.

As a consequence of the fifth duty, there should be adequate levels of supervision, including mentoring and monitoring of work in progress. This duty is owed to every employee as an individual, and especially to vulnerable people, including the young and those with disabilities, and remains even when the employee may be loaned to another employer.

2. NCC1 – Managing and controlling hazards in construction activities

Question 1

Construction workers using pneumatic drills can be exposed to hand–arm vibration during construction activities.

a **Identify** the:

 i daily vibration exposure action value; **(1)**
 ii daily vibration exposure limit value. **(1)**

b **Outline** possible health effects of exposure to hand–arm vibration. **(6)**

c **Outline** control measures to reduce the risks to workers from exposure to hand–arm vibration. **(12)**

Answer:

a i The daily vibration exposure action value is 2.5 m/s^2.

 ii The daily limit value is 5.0 m/s^2.

b Hand–arm vibration syndrome (HAVS) describes a group of diseases caused by the exposure of the hand and arm to external vibration. Some of these are described as Work-related Upper Limb Disorders (WRULDs), such as carpal tunnel syndrome and tenosynovitis. However, the best known disease is **vibration white finger (VWF)**, in which the circulation of the blood, particularly in the hands, is adversely affected by the vibration. The early symptoms are tingling and numbness felt in the fingers, usually some time after the end of the working shift. As exposure continues, the tips of the fingers go white and then the whole hand may become affected. This results in a loss of grip strength and manual dexterity. Attacks can be triggered by damp and/or cold conditions and, on warming, 'pins and needles' are experienced. If the condition is allowed to persist, more serious symptoms become apparent, including discoloration and enlargement of the fingers. In very advanced cases, gangrene can develop, leading to the amputation of the affected hand or finger.

The following control measures should be taken to reduce the risks associated with HAV:

▶ avoid, whenever possible, the need for vibration equipment. Consider whether the process could be automated to avoid the use of vibrating equipment, or, if this is not possible, selecting equipment producing lower levels of vibration;

▶ undertake a risk assessment which includes a soundly based estimate of the employees' exposure to vibration;

▶ develop a good maintenance regime for tools and machinery. This may involve ensuring that tools are regularly sharpened, worn components are replaced or engines are regularly tuned and adjusted;

> ▶ introduce a work pattern that reduces the time exposure to vibration, perhaps by job rotation or the provision of frequent breaks;
> ▶ issue employees with gloves and warm clothing. There is a debate as to whether anti-vibration gloves are really effective, but it is agreed that warm clothing helps with blood circulation, which reduces the risk of VWF. Care must be taken so that the tool does not cool the hand of the operator;
> ▶ introduce a reporting system for employees to use so that concerns and any symptoms can be recorded and investigated; and
> ▶ ensure that workers are given information, instruction and training in the control measures to be adopted.

It is important that drill bits and tools are kept sharp and used intact – an angle grinder with a chipped cutting disc will lead to a large increase in vibration as well as being dangerous. Finally, information and training for operators and supervisors are required by the Control of Vibration at Work Regulations.

Question 2

A senior manager is required to visit several different sites for a construction company and is required to travel long and short distances by road to visit the sites. The construction company is developing a policy to manage occupational road risk for its workers who are required to drive on the road at work.

a **Outline** the factors that increase the risk to the nominated person and other workers who are required to drive at work. **(10)**

b **Outline** factors that should be included within such a policy to manage occupational road risk. **(10)**

Answer:

a The factors that increase the risk to the nominated person and other workers who are required to drive at work may be considered under the headings of 'driver', 'vehicle' and 'journey'.

250

The factors that might increase the risk to the driver include competence, fitness or health condition and training. Competency issues include the validity of the driving licence, awareness of the company policy on work-related road safety and general driving experience. The driver's level of health and fitness should be sufficient for safe driving. Drivers who are most at risk should also undergo regular medicals. Poor eyesight is another significant risk for drivers. A recent survey has indicated that 25% of motorists have a level of eyesight below the legal standard for driving. All drivers should have regular (every 2 years) eyesight tests.

The vehicle will also affect the risks to the driver, and the factors concerning the vehicle are its suitability, its condition, with inspection and maintenance in mind, and the ergonomics of the driving position. Vehicles must be maintained in a safe and fit condition with regular checks on seat belts and head restraints. The load on lorries also presents a risk to the driver and should be stacked safely so that it cannot move during the journey. There must also be satisfactory arrangements for handling the load at either end of the journey. If the vehicle breaks down, the driver should know what to do to ensure his/her own safety.

Finally, the journey and route to be undertaken will also present risks to the driver because of the amount of traffic, the time of day when the journey is made and the sufficiency of the time allowed for its completion. The level of risk is also dependent on the type of roads being used, whether motorways, minor roads or those involving city centre congestion, and their condition. Route planning is crucial and schedules need to take sufficient account of busy periods on roads. Driving excessive distances without appropriate breaks considerably increases the risk to the driver, as do adverse weather conditions, such as snow, ice, heavy rain and high winds.

b An effective policy to manage occupational road risk should address legal requirements, driver training and health, monitoring driver performance and the number of recorded incidents. The policy should also cover vehicle and journey issues.

The legal responsibilities of the company include the determination of driving hours and their recording, statutory breaks and possible enforcement action by the authorities. The policy should cover the important and relevant aspects of the Highway Code; require that all speed limits be observed; and state that disciplinary action will be taken in the event of non-compliance with the agreed in-house rules and standards.

The policy should cover the need to ensure that all drivers must have a current and valid driving licence and that they know how to carry out routine safety checks such as those on lights, tyres and wheel fixings. Details of a prospective driver's previous driving experience should be examined and recorded. It should insist that drivers and passengers are adequately protected in the event of an incident. Crash helmets and protective clothing for those who ride motorcycles and other two-wheeled vehicles should be of the appropriate colour and standard. Drivers of Heavy Goods Vehicles (HGVs) must have the appropriate medical certificate. Staff should not drive, or undertake other duties, while taking a course of medicine that might impair their judgement.

The vehicle should also be covered by the policy and all vehicles must carry comprehensive insurance for use at work. All vehicles should be fit for purpose. New or replacement vehicles should be suitable for both driving and the health and safety of the public, and they should be regularly serviced and maintained. This is particularly important when pool cars are used, because pool car users often assume another user is checking on maintenance and the MOT. The safety-critical systems that need to be properly maintained are the brakes, steering and tyres. Similarly, if the car is leased and serviced by the leasing company, a system should be in place to confirm that servicing is being done to a reasonable standard. Each vehicle should have a handbook available for the driver and there should be a defect reporting system for drivers to use if they consider their vehicle is unsafe. Where privately owned vehicles are used for work, they should be insured for business use and, if necessary, have an appropriate MOT certificate test.

For lorries, the policy should address issues associated with the loads. If the load is hazardous, emergency procedures (and possibly equipment) must be in place and the driver trained in those procedures. The load should be stacked safely in the lorry so that it cannot move during the journey. There must also be satisfactory arrangements for handling the load at either end of the journey. The driver must be aware of the height of the vehicle, both laden and empty. Where tachographs are carried, they should be checked regularly to make sure that drivers are not driving for long periods without a break.

Finally, the policy should include topics associated with the journey and route. Drivers must be made aware of the dangers of fatigue caused by driving excessive distances without appropriate breaks. The recommendation of the Highway Code is for a 15-minute break every 2 hours. Motorways are the safest roads. Minor roads are suitable for cars, but they are less safe and could present difficulties for larger vehicles. Overhead restrictions, for example bridges and tunnels, and other hazards, such as level crossings, may present dangers for long and/or high vehicles, so route planning should take particular account of these. When planning journeys, sufficient consideration will need to be given to adverse weather conditions. No alcohol or recreational drugs must be consumed during the day of the journey until the journey is completed. Only minimal amounts of alcohol should be consumed on the day before a journey is to be made. Some prescribed and over-the-counter drugs and medicines can also affect driver awareness and speed of reaction. Drivers using such medicines must check with a doctor or pharmacist to ensure that it is safe to drive.

Question 3

Outline the factors to be included in a demolition method statement. **(8)**

Answer:

The contents of a demolition method statement will include the following:

▶ details of the method of demolition to be used, including the means of preventing premature collapse or the collapse of adjacent buildings and the safe removal of debris from upper levels to prevent overloading due to the debris;

▶ the age and previous use of the building, the type of construction used, such as steel frame or concrete beams, and the presence of underground voids;

▶ the stability of adjacent buildings and, where necessary, the use of temporary propping to prevent premature collapse;

▶ details of site access and security;

▶ details of the location of any underground or overhead services;

▶ details of arrangements for protection from falling materials;

▶ details of equipment, including access equipment, required and any hazardous substances to be used;

▶ arrangements for the protection of the public and the construction workforce against noise, and whether hazardous substances, such as asbestos, other dusts or lead, or flammable materials such as paints and/or solvents are likely to be released;

▶ details of the isolation methods for any services which may have been supplied to the site and any temporary services required on the site;

▶ details of personal protective equipment, such as hard hats, which must be worn by all personnel on site;

▶ first-aid, emergency and accident arrangements;

▶ training and welfare arrangements;

▶ arrangements for the separation, storage and disposal of waste;

▶ names of site foremen and those with responsibility for health and safety and the monitoring of the work;

▶ the coordination of all work activities on the site;

▶ the expected level of competence of site workers.

The site should be made secure with a perimeter fence and relevant signs posted to warn members of the public of the dangers. Further

protection of the workers and the public by creating exclusion zones may be needed.

Question 4

Lagging material that is suspected to contain asbestos has been disturbed during renovation work.

a **Outline** the actions that should be taken in order to protect the workers involved with the renovation work. **(6)**
b **Identify** the possible health risks from accidental exposure to the asbestos material. **(2)**

Answer:

a If asbestos is discovered during the performance of a contract, work should cease immediately and the employer should be informed. The area must be evacuated and signs erected indicating that no admittance is allowed. All ventilation and air conditioning systems must be shut down and a competent person should be engaged to assess whether asbestos is present, and its type if this proves to be the case. The workers should be informed of the suspicions and of the risks of entering the area. It would then be necessary to identify the individuals who might have been exposed and to explain the risks of exposure to them. A licensed contractor may need to be appointed for removal of the asbestos.

b Asbestos produces a fine fibrous dust of respirable dust size which can become lodged in the lungs. The fibres can be very sharp and hard, causing damage to the lining of the lungs over a period of many years. Asbestos is a possible chronic health risk. It can lead to one of the following diseases:

 ▶ asbestosis or fibrosis (scarring) of the lungs;
 ▶ lung cancer;
 ▶ mesothelioma – cancer of the lining of the lung or, in rarer cases, the abdominal cavity.

 Notes

 Notes

Notes

 Notes

Notes

T - #1073 - 101024 - C270 - 186/123/12 - PB - 9781138916791 - Matt Lamination